T0361840

Citrus

Production, Postharvest, Disease and Pest Management

Citrus

Production, Postharvest, Disease and Pest Management

S Mukhopadhyay

Cover photograph: *Courtesy Dr. Kishore Thapa, Officer-in-Charge, Citrus Dieback Research Station and Mr. Mahadev Chettri for providing the photographs of Darjeeling Mandarin.*

SCIENCE PUBLISHERS, INC.,
Post Office Box 699
Enfield, New Hampshire 03748
United States of America

Internet site: *http://www.scipub.net*

sales@scipub.net (marketing department)
editor@scipub.net (editorial department)
info@scipub.net (for all other enquiries)

Library of Congress Cataloging-in-Publication Data

Mukhopadhyay, S.
Citrus: production, postharvest, disease and pest management/S. Mukhopadhyay.
p. cm.
Includes bibliographical references (p.)
ISBN 1-57808-337-0
1. Citrus. I. Title.

SB369.M845 2004
634'.304--dc22 2003067339

All rights reserved, No part of this publication may be reproduced, stored in a retrieval system, or transmitted in any form or by any means, electronic, mechanical, photocopying or otherwise, without the prior permission of the copyright owner. Application for such permission, with a statement of the purpose and extent of the reproduction, should be addressed to the publisher.

ISBN 1-57808-337-0

© 2004, copyright reserved

Preface

Citrus is a very ancient crop known to occur over 4,000 years. Principal foci of this crop were in the Old World. It spreads to the New World with time. There are fascinating stories on its unique inter-country movement from ancient invasions to business routes. There is no doubt that *Citrus* has its worth. It is now the second most widely cultivated crop in the world (next to grape). Its cultivation extends from 40^0N latitude to 40^0S latitude. From the ancient time, its nutritional significance was well known particularly as the principal source of vitamin C and Folic acid. Now its several therapeutic properties have also been established particularly for its pectins and flavonoids. Grapefruit pectin can significantly lower the plasma cholesterol levels and produce a significant improvement in LDL/HDL ratio. Citrus flavonoid tangeretin has the ability to prevent invasion of normal tissues by cancer cells. Citrus hesperidin, naringin, tangeretin and nobiletin have ant-inflammatory and anti-allergic activities. These flavonoids also improve circulatory system. Its therapeutic and nutritive values along with its taste and flavor have placed it in the regular dietary list of the people living in advanced countries.

International scientists also are paying a special attention to this crop. It is very interesting to note that three International Organizations, namely, International Society of Citriculture, International Society of Citrus Nurserymen and International Organization of Citrus Virologists, regularly hold Congresses to discuss this crop.

There are also voluminous publications on this crop. The University of California, CA, USA published six volumes on "Citrus Industry" in 1967. CABI, Willingford, UK and Cambridge University Press, Cambridge made brief publication on "Citrus" and "Biology of Citrus" in 1994 and 1996 respectively. CIRAD, Montpellier, France and GTZ, Eschborn, Germany jointly published a practical guideline for citriculture in 1998. American Phytopathological Society published a book on " Management of Citrus Health" in 1999. But "*Citrus*" is to be seen in a holistic way integrating "Production" and "Utilization". This book aims at this approach.

First chapter of this book deals with taxonomy and the genetic diversity of citrus as the knowledge on this subject is imperative for it's understanding and developing improvement programs. Second chapter describes commercial cultivars developed in nature or human endeavor cultivable in diverse biotic and a-biotic conditions for improved commercial use. Third chapter provides information on rootstocks holding the key for qualitative and quantitative improvement of yield and use, and also wide adaptability of the crop. Fourth chapter describes a unique feature of this crop, the plyembryony in seeds. Fifth chapter on "Production Technology" is self explanatory, a chapter on which a potential grower can rely upon. Production has been integrated with commercial use in the sixth chapter on "Postharvest Technology".

As threats from diseases and pest increase with the development of new cultivars, chemical farming, and reliance on organic pesticides, they have been dealt with in separate chapters. It is a nascent effort to present "*Citrus*" in its totality.

August 2003 S. Mukhopadhyay

Contents

List of Tables

List of Tables

List of Figures and Plates

1

Taxonomy, Chromosomal Organization and Genetic Diversity

Citrus is a genus of the plant family Rutaceae. This family is divided into 7 subfamilies comprising 11 tribes and 93 genera (Engler, 1931). *Citrus* belongs to the subfamily Aurantioideae or the Orange subfamily. It contains 2 tribes and 6 subtribes comprising 33 genera and over 215 species. Most members of this subfamily have white, fragrant flowers and gland-dotted aromatic fruits. The genus *Citrus* is the most specialized, having juicy fruits with pulp vesicles or juicy sacs. Swingle first classified the species of *Citrus* in 1943 and again in 1967. He recognized 16 species and 8 botanical varieties of *Citrus*. On the contrary, Tanaka recognized 159 species (Tanaka, 1954, 1969). It has been argued (Jones, 1990) that most of the species described by Tanaka are hybrids or mutants.

TAXONOMIC CHARACTERISTICS OF *CITRUS*

Citrus L. is a deciduous to evergreen tree or shrub with sharp spines. Leaves are unifoliate, alternate, coriaceous or curtaceous and punctate with aromatic pellucid glands. Petioles winged or wingless, articulated with leaf blades. Flowers solitary, in cymes or racemes, small or large, bisexual or staminate, sweet scented. Calyx 4–5-lobed, cup-shaped, 4–8 (normally 5) in number. Petals white or tinged, coriaceous, linear, gland-dotted; its shape is imbricate. There are 20–60 polyadelphous stamens. Ovary free, filamentous at base, many-celled (10–14) with terete; there are two series of deciduous ovules in each cell. Fruit a segmented hesperidia containing seeds near the ventral side and stalked, fusiform; pulp vesicles contain sweet or sour juice. Rind is foveolated with oil glands, turns yellow or orange or red at full maturity. There may be no seed or there may be many obovoid to flattened-ovoboid seeds containing one or more white or green embryo(s). Flowers are cross-pollinated and natural hybrids arise with time. Bud mutations also occur, producing different morphological features in different branches of a particular plant.

SPECIES OF *CITRUS* AND THEIR DISTRIBUTION

Jones (1990) divided *Citrus* into 2 subgenera, *Citrus* and *Papeda,* on the basis of the character of petiole, flower, stamen, and pulp vesicle. In *Citrus,* the petioles are wingless or narrowly winged, flowers large and fragrant, stamens cohering, and pulp vesicles lacking acrid oil droplets. *Papeda,* on the other hand, has large and broadly winged petiole, small flower, free stamens, and pulp vesicles contain numerous acrid oil droplets. Jones placed 10 species under *Citrus* and 6 species under *Papeda.* The species, varieties, distribution, status, and uses of *Citrus* as outlined by him are given in Table 1.1.

Species outlined within the subgenus *Papeda* include *C. ichangenesis* (ichangpapeda), *C. latipes* (khasi papeda), *C. micrantha* var. *microcarpa* (small-flowered papeda), *C. celebica* (Celebes papeda), *C. macroptera* var. *kerrii,* var. *annamensi* and *C. hystrix* (Mauritius papeda). *C. ichangenesis* is distributed in west, central and southwest China, occurs wild, cold tolerant and mostly used for hybridization. *C. latipes* is distributed in northeast India and northern Burma, occurs wild, and used for hybridization and as rootstock. *C. micrantha* var. *microcarpa* is distributed in the southern Philippines, sparingly cultivated and used for medicinal purposes (hair wash). *C. celebica* is distributed in northern Sulawesi, and southern Philippines, occurs wild, sparingly cultivated, and used for hybridization. *C. macroptera* is distributed in Indonesia, Malaysia, Papua New Guinea, and Thailand, occurs wild, and used for hair wash, domestic rootstock, and hybridization. *C. hystrix* is distributed in Burma, western Malaysia and Philippines, cultivated, widely naturalized, and used for flavoring, hair wash, wood, and medicinal purposes.

Domestication of *Citrus* began in fact in Southeast Asia (Northeastern India, and neighboring regions of Burma and China) more than four thousand years ago. This region is the primary center of origin of different *Citrus* species. Mediterranean and Caribbean regions are the areas of secondary diversification. The most ancient *Citrus* species, citron, was introduced to Mesopotamia, Greece, and Palestine during 600, 300, and 136 years B.C. respectively. From this species new types emerged over the centuries (*C. aurantifolia, C. limon, C. aurantium, C. sinensis*). Some of these types were introduced into Japan during the 1st century, and to Genoa, and Portugal during the 10th and 13th century respectively. Two other ancient species are *C. grandis* and *C. reticulata. C. grandis* originated in Malaysia, Indonesia, and Vietnam. *C. reticulata* originated in southern China and Japan. Mandarins were introduced in the Mediterranean basin during the 19th century. From this basin, different *Citrus* species were taken to the Caribbean region. From this region, again, they were taken to Mexico during the early 16th century, Brazil during the middle of the

Table 1.1: Species, varieties, distribution, status and uses of *Citrus* and *Papeda* (Jones, 1990)

Species (common name): *C. medica* (citron)
Variety: var. *sarcodactylis,* var. *ethrog*
Distribution: eastern India to Upper Burma
Status: sparingly cultivated
Uses: edible, medicinal, perfumery, ornamental, ceremonial

Species (common name): *C. limon* (lemon)
Distribution: eastern India to Upper Burma
Status: cultivated
Uses: essential oil, citric acid, hybridization

Species (common name): *C. aurantifolia* (lime)
Distribution: western Malaysia
Status: cultivated
Uses: edible, medicinal, flavoring, essential oil, citric acid, hybridization

Species (common name): *C. aurantium* (sour orange)
Variety: var. *myrtifolia* (myrtle-leaf orange)
Distribution: eastern India to Upper Burma
Status: cultivated, naturalized in some area
Uses: edible, medicinal, flavoring, perfumery, liqueur, hybridization

Species (common name): *C. sinensis* (sweet orange)
Distribution: northeastern India, Upper Burma, Southern China
Status: cultivated
Uses: edible, medicinal, flavoring, essential oil, fodder, rootstock, hybridization

Species (common name): *C. reticulata* (mandarin orange)
Variety: var. *austera* (sour mandarin)
Distribution: China
Status: cultivated
Uses: edible, medicinal, rootstock, hybridization

Species (common name): *C. maxima* (pomelo)
Variety: var. *racemosa* (grapefruit)
Distribution: Thailand, Malaysia
Status: cultivated, occasionally wild
Uses. edible, medicinal, hybridization

Species (common name): *C. indica* (wild Indian orange)
Distribution: eastern Himalayas, Assam
Status: wild

Species (common name): *C. tachibana* (tachibana)
Distribution: Taiwan, Japan
Status: wild and semiwild
Uses: rootstock

Species (common name): *C. halimii* (sultan lemon)
Distribution: Malaysia, Thailand
Status: wild

20th century, the USA during the late 16th century, Peru during the early 17th century, and Australia during the late 18th century. *Citrus* species were carried from the Mediterranean basin to Arabia, Mozambique during the 10th century, and South Africa during the 17th century (Ollitrault and Luro, 2001)

CHROMOSOMAL ORGANIZATION AND GENETIC DIVERSITY OF *CITRUS*

Citrus plants in nature are normally diploid with chromosome number 2n=18. However, the flowers are cross-pollinated and produce natural hybrids. It has been observed that the ploidy level in nature may be as high as octaploids. Bud mutations also occur, leading to different morphological features in different branches of the same plant. So, there is vast scope for wide chromosomal diversity and so-called biodiversity.

Citrus biodiversity was first documented in China as early as 1178 when Hanyienchi described 27 *Citrus* cultivars in that country (Zhang et al., 1992). During the middle of the 20th century, Swingle (1943) and Tanaka (1915–30) conducted extensive exploration on citrus germplasm around the world. They published their observations in 1943 and 1967, and 1954 and 1969 respectively. Hodgson (1937), while living in India, described different types fruits in *Citrus* species found in that country. Swingle recognized 16 species whereas Tanaka recognized 159 species. Subsequent workers found that many species recognized by Tanaka were actually hybrids and mutants. Scora (1975) reviewed the history of *Citrus*.

Recent analyses of the data available on *Citrus* species using advanced methods such as numerical taxonomy and chemical taxonomy, etc., indicate that all species of *Citrus* center around only three basic species or biotypes: *C. grandis* (pomelo), *C. medica* (citron), and *C. reticulata* (mandarin); others evolved through natural hybridization or mutation. Lime and lemon are considered trihybrids involving citron, pomelo and Microcitrus; sour and sweet oranges are hybrids between mandarin and pummelo. Barrett and Rhodes (1976) illustrated the evolution of different commercial citrus varieties from the three basic species (Figure 1.1).

The diversity of commercial varieties of Citrus became a concern of citriculturists for variety development and conservation. Giacometti and van Sloten (1984) described three gene pools of cultivated *Citrus* species: primary, secondary, and tertiary. The primary pool involves *C. aurantifolia, C. grandis, C. limon, C. medica, C. grandis* var. *racemosa, C. madurensis,* and all the cultivars of sweet oranges and mandarins (including tangerines). The secondary pool involves *C. macroptera, C. hystrix, C. ichangensis,* all the species of *Fortunella* and *Poncirus trifoliata*; the tertiary gene pool consists of all the species of other genera closely related to

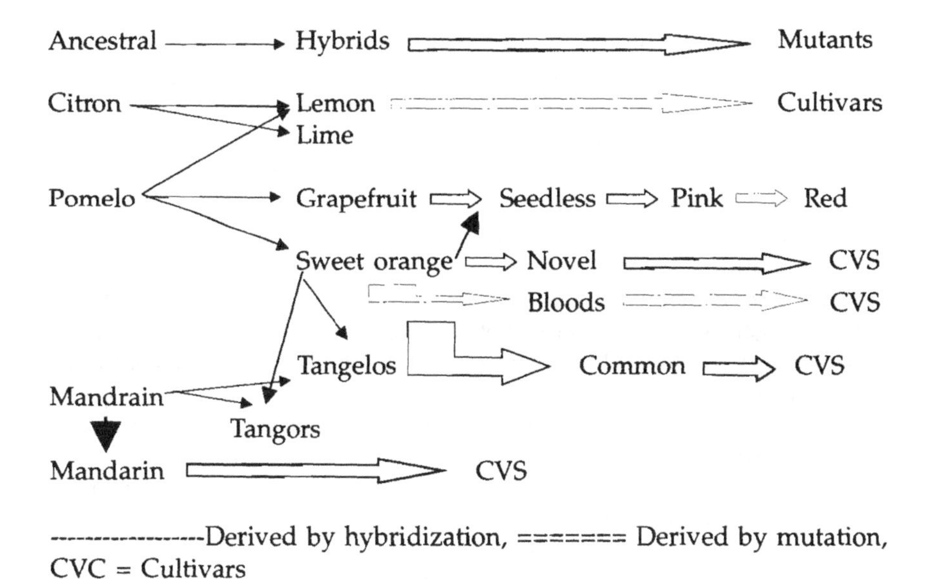

Figure 1.1: Probable origin of *Citrus* cultivars through hybridization and mutation.

Citrus, some of which may be able to hybridize *Microcitrus, Clymenia,* and *Eremocitrus.* Chromosomal studies on the common cultivated species also show interesting features in their natural diversity:

1) *C. medica*: lower allelic diversity that is indicative of a strong homozygosity and the lack of intercultivar diversity;
2) *C. paradisi, C. sinensis,* and *C. aurantium*: moderate heterozygosity and allelic diversity; intercultivar diversity nonexistent;
3) *C. limon*: important allelic diversity due to very high heterozygosity;
4) *C. latifolia* and *C. aurantifolia*: high heterozygosity and medium intercultivar diversity;
5) *C. grandis*: higher genotypic and allelic diversity combined with moderate heterozygosity;
6) *C. reticulata*: low heterozygosity combined with important allelic diversity due to high genotypic diversity.

All types of biotechnological methods have been applied to differentiate the gene maps of scions and rootstocks of commercial varieties but the gene pool or mapping of wild *Citrus* still remains inadequately assessed. Undoubtedly during the course of evolution, the germplasm of many wild species have become part of several cultivated species, some of which virtually remained unchanged in some forms of lime (*C. aurantifolia*), pomelo (*C. grandis*), citron (*C. medica*), limau-burut (*C. hystrix*); other cultivars are either hybrids or in some cases, long-persisting clonal derivatives of mutated hybrids. It is feared that the Citrus biodiversity in their homelands has been lost as it is now difficult to find true primary forests with such biodiversity, but possibly wild forms may be found in different forests, for example wild *C. macroptera* occurs in Borneo, Philippines, Celebes, New Guinea, and perhaps in New Caledonia. It is further conjectured that in many places certain species of *Citrus* have perhaps become thoroughly naturalized and thrive as distinct species in the reconstituted secondary forests of those places, which may be the case in parts of the West Indies and certain South American countries where the genus was originally exotic. It is believed that hybrids of *Citrus* in which a percentage of wild germplasm content is probably common, occur mostly in Southeast Asia. With the loss of tropical and subtropical forests in this region, such germplasms have been either lost or still exist at a very vulnerable state.

But the potential of the secondary diversification exits through natural and artificial hybridization. Genetic exchange among the basic taxa is limited; recombination between chromosomes of different taxa is also restricted due to differential in size of the nuclear genomes. But limited gene flow is often affected by polyembryony and relative compatibility to hybridization. Thus secondary diversities appear through intraspecific and intergeneric hybridization.

CLOSELY OR DISTANTLY RELATED SPECIES OF *CITRUS* OF ECONOMIC IMPORTANCE

Swingle (1967) recognized two tribes under Aurantioideae, Clauseneae (very remote and remote citroids), and Citreae (citrus and citroids). Clauseneae has been divided into three subtribes, namely, Micromelinae (very remote citroids), Clauseninae (remote citroids) and Meriliinae (large fruited remote citroids). Tribe Citreae has been divided into two subtribes, namely, Triphasiinae (minor citroids) and Citrinae (citrus). Members of Citrinae are often used in breeding programs for citrus improvement. This subtribe has been divided into three groups, Primitive Citrus with five genera (*Severinia,* distributed in India to China and Malaysia; *Pleiospermium,* distributed in India, Indonesia, Malaysia, Vietnam; *Barkillanthus,* distributed in Indonesia, Malaysia; *Limnocitrus,* distributed in Indonesia, Vietnam; and *Hesperethusa,* distributed in India to Thailand). Near Citrus with one genus, *Atalan*tia, is distributed in India to Indochina, Malaysia and Indonesia. True Citrus includes commercial citrus. This Group includes four genera, many species of which have been used for breeding purposes. Important features of these genera are:

Genus *Fortunella:* Four species, distributed in Southern China to Indochina, mostly cultivated, fruits are edible and plant used for medicinal, ornamental, and hybridization purposes. It has cold tolerance.

Genus *Clymenia*: Two species, distributed in Indonesia, Papua New Guinea, mostly wild, used for ornamental purposes. Fruits edible.

Genus *Microcitrus*: Seven species, distributed in Papua New Guinea and Australia, mostly wild, used for ornamental purposes. Fruits edible.

Genus *Citrus*: Divided into two subgenera, *Citrus* and *Papeda*. Most of the species are commercial. Their distribution and uses have already been described. Characteristics of the commercial species and cultivars dealt with in the next chapter.

MOLECULAR APPROACH TO CITRUS TAXONOMY, GENETICS AND BIODIVERSITY

The taxonomy of citrus has been reviewed in the last two decades in the light of the current molecular technology. Torres et al. (1978), Green et al. (1986), Handa et al. (1986), Roose (1988), Ollitrault and Faure (1992) and several other workers applied isozymes and RFLPs to study citrus biodiversity. Their observations supported the ancestral status of the three *Citrus* species described earlier (*Citrus medica, Citrus grandis*, and *Citrus reticulata*). Luro et al. (1992) applied Random Amplified Polymorphic DNA (RAPD) to study citrus genetics and taxonomy. They studied 18 different species belonging to three genera (*Citrus, Fortunella,* and *Poncirus*). Through their studies they could clearly separate the ancestral groups. According to them *C. aurantifolia* is probably a trihybrid involving *C. medica, C. grandis* and *Microcitrus*. Their observations confirmed the views of Barrett and Rhodes (1976). They found *C. aurantium* to be of *C. reticulata* origin with some genes of *C. grandis. C. limon* they suspected to be a trihybrid of *C. medica, C. aurantifolia* and an unidentified species. *C. sinensis* was found to be closely related to *C. reticulata* and distantly related to *C. grandis,* confirming the view of Tatum et al. (1974). They suspected *C. paradisi* to be a hybrid between *C. sinensis* and *C. grandis,* as presumed by Swingle (1943), Robinson (1952) and Albach and Redman (1969). They speculated the origin of *C. madurensis* (Calamondin) and *C. jambhiri* (Rough lemon) from *C. reticulata–Fortunella* and *C. reticulata–C. medica* respectively, but they could find no evidence for the *Fortunella* background of *C. madurensis. C. halimi* in their study behaved as an outgroup, indicating its wide divergence from cultivated species.

Motohashi et al. (1992) performed DNA fingerprinting in 25 citrus cultivars mostly belonging to *C. unshiu* Marc, *C. nobilis* Loureiro, *C. delicosa* Tenore, *C. paradisi* Macf, *C. tangerina* hort. ex Tanaka, and *C.*

clementine hort. ex Tanaka. They found DNA fingerprints to be very useful for differentiating cultivars at the molecular level and this technique applied correctly could effectively solve the problems in the citrus taxonomy.

Asins et al. (1998) studied the genetic diversity contained in the citrus germplasm bank IVIA (Valencia, Spain). It consisted of 198 cultivars and accessions of 54 species of *Citrus* and 13 related genera of Aurantioideae subfamily. They used molecular markers to study the genetic diversity. The lowest genotypic variability was found to occur in *C. myrtifolia, C. deliciosa* (Willow leaf mandarin), *C. paradisi, C. limon,* and *C. sinensis* while the highest value was found in *Severinia buxifolia.* All *Citrus* species are not highly heterozygous but a broad spectrum of heterozygosity is found in many of them. Lemons, limes, and *C. begamia* show a very high percentage of heterozygosity, indicating their origin through interspecific hybridization. Intraspecific variability is very limited because of the apomictic reproduction. In addition, some species are self-pollinating with low heterozygosity.

Citrus has two clearly defined principal groups: orange-mandarin, and lime-lemon-citron. *Microcitrus* is closer to *Citrus* than to *Fortunella.* The former is related to the citron-lemon group, and the latter to the orange-mandarin group. *Poncirus* is far removed from the rest of the genera. *Atalantia* species are also very distantly located and highly variable.

Orange-mandarin group is divided into three subgroups: sour orange (*C. clementina, C. tangerina, C. nobilis, C. myrtifolia, C. aurantium*), sweet orange (*C. sinensis, C. temple, C. unshui*), and a third subgroup containing the rest of the mandarin species. Orange-fruited *Citrus* species form a compact group. It may be that all mandarins do not fall under a single species, *C. reticulata.* Citron, pomelo, and ancient lemon subgroups form a cluster to which the species belonging to the subgenus *Papeda* and the cultivated limes, lemons, and bergamots are related. Thus the study with respect to the dendrogram of Aurantioideae based on the chord distance and neighbor-joining clustering method conducted by these workers suggest wide genetic differences among different genera of citrus and the wide extent of adaptation in the course of evolution of the subfamily.

Current studies conducted by biotechnological methods on the phylogeny, diversity, genetics, and chromosomes have unfolded some very interesting features of this historical genus. Nicolosi et al. (2000) using molecular markers, made interesting observations on the phylogeny and origin of different *Citrus* and related species. They took 36 accessions belonging to *Citrus* together with one accession from each of the related genera *Poncirus, Fortunella, Microcitrus,* and *Eremicitrus.* Phylogenetic analysis conducted by different methods (262 RAPDs, 14 SCARs) indicated the closeness of *Fortunella* to *Citrus* while the other three related genera appeared distant from *Citrus* and each other. They also observed closeness between *Citrus* and *Papeda* except for *C. celebica* and *C. indica*.

In contrast to the earlier observations on the existence of two groupings, Nicolosi et al. (2000) found three clusters, each containing one genotype specific for the true species. No confirmation of the subgeneric status of *Citrus* and *Papeda* was obtained. The distantness of *C. medica* and *C. indica* from other species of *Citrus* and related genera was established. They found *C. ichangensis* to be the ancestral species of the mandarin cluster including *C. tachibana*. The "*Pomelo cluster*" included lemon and Palestine sweet lime, led by *C. latipes*. *C. aurantifolia* was found to be located in the "*Micrantha cluster*".

These authors also confirmed the origin of lemon and Palestine sweet lime, bergamot, and volkamer lemon as hybrids of citron and sour orange while Rangpur lime and Rough lemon were found to be hybrids of citron and mandarin. According to them the Mexican lime originated as the hybrid of *C. micrantha* (female parent) and *C. medica* (male parent).

Guerra et al. (2000) took a different approach to the study of the phylogeny and genetic relationship between various *Citrus* and related species. Using molecular markers they studied chromosome and heterochromatin banding patterns in 17 species belonging to 15 genera of the Aurantioideae with fluochromes chromacin (CMA) and 4′-6′-diamidino-2-phenylindole-2-HCl (DAPI). They found all species to be diploids except one tetraploid (*Clausena excavata*) and two hexaploids (*Glycosmis parviflora* agg. and *G. pentaphylla* agg.). When ct DNA (chloroplast DNA) sequence data was used as a phylogenetic background, generally more basal genera exhibited very small amounts of heterochromatin (*Glycosmis, Severinia, Swinglea,* etc.), whereas relatively advanced genera displayed more CMA^+ bands (*Merrillia, Ferniella, Fortunella*). These bands were located on the larger chromosomes and at the telemetric regions of larger arms. It appears that one of the largest chromosome pairs has been conserved throughout the subfamily with very little heterochromatin. The heterochromatin-rich patterns observed in different clades of Aurantioideae appear quite similar. These observations suggest the monophyletic nature of the core group of Citrae with *Ermocitrus, Microcitrus, Clymenia, Poncirus, Fortunella,* and *Citrus.*

Baril (2001b) evaluated the genetic relationships in *Citrus* by means of sequence tagged micro satellites (STMS); 73 individuals of *C. reticulata, C. sinensis, C. aurantium, C. paradisi, C. maxima, C. limon, C. aurantifolia* and *C. medica* were analyzed by *simple sequence repeats* (SSRs) and by amplification of their 10 loci. He further evaluated the potentiality of STMS for cultivar identification within species performing neighbor-joining analysis. On the basis of the dissimilarity matrix obtained from the different genotypes the author obtained genetic aggregations in which three groups could be clearly identified, viz. *orange-mandarin, pomelo-grapefruit,* and *lime-lemon-citron* group.

Goldschmidt (2000) studied the phylogeny of *Citrus* and related genera by assessing putatively hybrid taxa using molecular markers (RFLPs). He studied 73 accessions from 45 *Citrus* species and 12 accessions from related genera. The data were programed in a computer and could fully account for *C. bergamia* by *C. aurantium* *C. limetta*. The RFLP profile of *C. limonia* was fully accounted for by 6 mandarin species *C. aurantifolia* or *C. macrophylla,* while 5 mandarins *C. medica* accounted for all but one *C. limonia* band. But the datasets could not define the ancestors of *C. jambhiri.*

Samuel et al. (2001) confirmed the monophyletic origin of subfamily Aurantioideae by sequencing plastid DNA from 23 genera and 47 species. But they could not justify the inclusion of *Murraya* and *Merrillia* under tribe Clausenae as they fit close to Citrinae.

Matsuyama et al. (2001) observed specific *dispersed repetitive sequences* of citrus genomic DNA (MCS-26a, JA-5, and JB-7) to the families, which proved highly variable among the members. The distribution of MCS 26a varies widely among members of Aurantioideae. They confirmed the Rutaceae-specific repeated sequences (MCA26a, JA-5 and JA-7) using standard slot blot hybridization. They found wide distribution of MCA 26a in subfamily Aurantioideae whereas the extent of distribution of JA-5 amplification varied at the generic level. Gulsen and Roose (2001) measured *intersimple sequence repeats* (ISSRs), *simple sequence repeats* (SSRs), and *isozymes* and established the existing concept on genetic diversity and phylogeny. They also confirmed the ancestral nature of pomelo, citron, and mandarin.

Baril (2001a) assessed the genetic relationships between 22 citrus cultivars (sweet oranges, lemons, grapefruits, clementines, and several other mandarin biotypes) by isozymes and RAPD markers. 4 isozyme systems (GOT, IDH, PMG, and PGT) used by him revealed 7 loci, of which 4 were polymorphic. These systems discriminated sweet oranges, lemons, grapefruits and clementines at specific levels. He found a specific isozyme pattern in each of the mandarin cultivars with the exception of Carvalhais and Fremont. The RAPD analysis using 36 decamer primers revealed 289 markers, of which 48.4% were polymorphic. This technique proved able to discriminate among all the species and to distinguish among the mandarin cultivars, including the aforesaid two, but unable to discriminate the cultivars of other types.

Ollitrault et al. (2003) determined the gene size of the genomes of different genotypes. They found small genomes in diploid genotypes (0.73–0.82 pg DNA). In triploid genotypes the size of the DNA was 1.17 pg, as found in four cultivars of lime (Tahiti, Bears, Elkseur, and IACSRA 618).

Befu et al. (2001) attempted groupings of different *Citrus* species by studying their chromosomes. They found 5 types of chromosomes (A, B, C, D, E) in pomelo, citron, mandarin and grapefruit. They found D and E as basic types for *Citrus*. Other types varied quantitatively in different species.

Mukhopadhyay and Thapa (2001) looked into *Citrus* groupings by statistical methods. They conducted cluster analysis using 30 numerical characters of the collections of indigenous citrus germplasm occurring in forests of the eastern Himalayas. They could cluster 14 collections into 5 groups, indicating the genetic diversity of these species (Table 1.2).

Table 1.2: Cluster of 14 different wild/semiwild *citrus* species and their biotypes collected from forests around Kalimpong hills of Darjeeling district, West Bengal, India (Mukhopadhyay and Thapa, 2001)

Cluster	*Citrus* species and their biotypes
I	*C. grandis*-I, *C. grandis*-II, *C. limonia, C. jambhiri*
II	*C. medica*-II, *C. aurantifolia*
III	*C. grandis*-III, *C. aurantium*-I, *C. aurantium*-II
IV	*C. medica*-III, *C. indica*
V	*C. medica*-I, *C. medica*-IV, *C. medica*-V

REFERENCES

Albach RF and Redman GH (1969). Composition and inheritance of flavones in Citrus fruit. *Phytochemistry* 8: 127–143.

Asins MJ, Mestre PF, Herrero R, Navarro L, and Carbonell EA (1998). Molecular markers: a continuously growing biotechnology area to help citrus improvement. *Fruits* 53 (5): 293–302.

Baril C (2001a). Assessment of the genetic relationships among *Citrus* species and varieties by isozyme and RAPD. Proc. Int. Symp. Molecular Markers for Characterizing Genotypes and Identifying Cultivars in Horticulture (eds. L Cabrita, P Elisiario, J Laitao, A Guerreiro and C Dore). *Acta Horticulturae* 546: 177–181.

Baril C (2001b). Evaluation of genetic relationship in *Citrus* genus by means of sequence tagged microsatellites. Proc. Int. Symp. Molecular Markers for Characterizing Genotypes and Identifying Cultivars in Horticulture (eds. F Luro, D Rist, P Ollitrault, and C Dore). *Acta Holticulturae* 546: 237–242.

Barrett HC, and Rhodes AM (1976). A numerical taxonomic study of affinity relationships in cultivated *Citrus* and its close relatives. *Syst. Bot* 1: 105–136.

Befu M, Kitajima A, and Hasegawa (2001). Chromosome composition of *Citrus* species and cultivars based on the chromycin A3 (CMA) banding patterns. *J. Jpn. Soc. Hort. Sci.* 70 (1): 83–88.

Engler A (1932). Rutaceae *In*: Die Naturilichen Pflanzen amilien (eds. A Engler and K Prantl) Band 19a Verlag von Wilhem, Engelmann, Leipzig, pp. 187–357.

Giacometti DC and van Sloten DH (1984). Citrus Genetic Resources. *Proc. Int. Soc. Citriculture*. Int. Citrus Cong., Sao Paulo, Brazil, vol. 1, pp. 1–3.

Goldschmidt EE (2000). RFLP analysis of the origin of *Citrus bergamia, Citrus jambhiri* and *Citrus limonia*. Proc.1st Int. Symp. Citrus Biotech. (eds. CT Federeci, ML Roose, RW Scora and R Goren). *Acta Horticulturae,* 535: 55–62.

Green RM, Vardi A and Galun E (1986). The Palstone of *Citrus*. Physical map, variation among Citrus cultivars and species, and comparison with related genera. *Theor. Appl. Genet.* 72: 761–769.

Guerra M, Santos KGB-dos, Barros-e-Silva AE, Ehremdorfer M, and dos-Santos KGB (2000) Heterochromatin banding patterns in Rutaceae–Aurantioidae—a case of parallel chromosome evolution. *Am. J. Botany* 87 (5): 735–747.

Gulsen O and Roose ML (2001). Lemons: diversity and relationships with selected *Citrus* genotypes as measured with nuclear genome markers. *J. Am. Hortic. Sci.* 126 (3): 309–317.

Handa T, Ishzawa Y, and Ogaki C (1986). Phylogenetic study of Fraction I protein in the genus *Citrus* and its closely related genera. *Jpn. J. Genet.* 61: 15–24.

Hodgson RW (1937). The citrus fruits of India. *Calif. Citrograph* 22: 504, 513–514, 517.

Jones DT (1990). A background for the utilization of Citrus genetic resources in Southeast Asia. I: Classification of Aurantioidae. *In*: 4th Int. Asia Pacific Conf. Citrus Rehabilitation (eds. B Aubert, S Tonyporn and D Buangsuwon). Chiang Mai, Thailand, FAO-UNDP RAS/86/022 Regional Projects, pp. 31–37.

Luro F, Laigret F, Bove JM, and Ollitrault P (1992). Amplification of random amplified polymorphic DNA (RAPD) to *Citrus* genetics and taxonomy. Proc. Int. Soc. Citriculture VII Int. Citrus Cong. Acireale, Italy, pp. 225–228.

Matsuyama T, Omura M, and Akihama T (2001). Distribution of Rutaceae-specific repeated sequences isolated from *Citrus* genomes. *Ann. Botany* 87 (6): 845–849.

Mukhopadhyay S and Thapa K (2001). Citrus legacy and biodiversity in Eastern India. *Sci. Culture* 67 (1–2): 42–46.

Nicolosi E, Deng ZN, Gentile A, Malfa-S-la, Continella G, Tribulato E, and la-Falfa S (2000). Citrus phylogeny and genetic origin of important species as investigated by molecular markers. *Theor. Appl. Genet.* 100 (8): 1155–1166.

Ollitrault P and Faure X (1992). Systeme de reproduction et organization de la diversite' gene'tique dans le genre Citrus. Proc. Int. Col. "Complexe d'especes, flux de genes et resources gene'tiques des plantes". BRG, Paris, pp. 135–151.

Ollitrault P and Luro F (2001). Citrus. *In*: Tropical Plant Breeding (eds. A Charrier et al.). Science Publ. Inc., Enfield NH, USA, pp. 55–77.

Ollitrault P, Jacquemond C, Dubois C, and Luro F (2003). *In*: Diversity of Cultivated Tropical Plants (eds. P Hamon et al.). Science Publ. Inc., Enfield, NH, USA, pp. 206–227.

Robinson TR (1952). Grapefruit and Pomelo. *Econ. Bot.* 6: 228–245.

Roose ML (1988). Isozymes and DNA restriction fragment length polymorphism in *Citrus* breeding and systematic. Proc. Int. Soc. Citriculture (eds. Gorel and Mendel). VI Citrus Cong. Philadelphia, PA, pp. 155–165.

Samuel R, Ehrendorfer F, Chase MW and Greger H (2001). Phylogenetic analyses of Aurantioideae (Rutaceae) based on non-coding plastid DNA sequences and phytochemical features. *Plant Biology* 3 (1): 77–87.

Scora RW (1995). On the history and origin of *Citrus*. *Bull. Torrey Bot. Club* 102: 369–375.

Scora RW (1988). Biochemistry, taxonomy and evolution of modern cultivated citrus. *In*: VI Int. Citrus Cong. pp. 277–287.

Swingle TW (1943). Botany of *Citrus* and its wild relatives of the Orange subfamily *In*: Citrus Industry (eds. HJ Webber and LD Bachelor). Univ. Calif. Press, Berkeley, CA, vol 1, pp. 129–474.

Swingle TW (1967). Botany of *Citrus* and its wild relatives of the Orange subfamily *In*: Citrus Industry (eds. W Reuther, H Webber and L Bachelor). Univ. Calif. Press, Berkeley, CA, vol. 1, pp. 190–430 (rev. ed.).

Tanaka T (1954). Species problems in *Citrus*. Jpn. Soc. Promotion of Science, Uemo, Tokyo, pp. 152.

Tanaka T (1969). Fundamental discussion of citrus classification. *Studia Citrologia* 14: 1–6.

Tatum JH, Berry RE, and Hearn CJ (1974). Characterization of Citrus cultivars and separation of nucellar and zygotic seedlings by thin layer chromatography. *Proc. Florida State Hort. Soc.* 87: 75–81.

Torres AM, Soost RK, and Diedenhopen U (1978). Leaf isozymes and genetic markers in *Citrus. Am. J. Bot.* 65: 869–881.

Zhang WC, Deng XX, and Deng ZA (1992). Citrus germplasm preservation and varietal improvement works in China. *Proc. Int. Soc. Citriculture,* 1: 67–71.

2

Commercial Citrus Varieties and Hybrids

Citrus varieties and cultivars that are now widely cultivated include mandarins, sweet oranges, lemons, limes, pomelos, and grapefruits. At the global level the relative production of different citrus varieties is 18.8%, 69.7%, 5.3% (lime and lemon), and 6.7% (pomelo and grapefruit) respectively (Timmer and Duncan, 1999).

In India, mandarins constitute maximum area and production (approximately 45%). Sweet oranges come next and constitute about 30% of the total citrus fruits produced in this country, followed by acid lime and lemon. Pomelo and grapefruit are not of commercial significance at present. Taxonomic characters of different commercial citrus fruits are given below:

A. MANDARIN (*CITRUS RETICULATA* BLANCO)

The mandarin tree is normally a small to large, spiny, with a dense canopy of slender branches. Leaves lanceolate, with narrowly winged petiole. Flowers are white and perfect. Fruit depressed, globose, yellow or reddish-orange when mature, with loose skin, hollow core, and thin easily separable rind. Pulp segmented, containing 10–14 separable segments, and 12–20 small beaked seeds. These are clavate, with fin-like projection of testa, and brown chalazal cap. Each seed normally measures 11–13 mm in length and 5–7 mm in breadth. Surface of the seed smooth, veined, and cream colored. Cotyledons green.

1. Mandarins of Different Countries

In the mandarin-growing countries of the world, the species/cultivars have been taken from the countries of origin, naturalized, and/or raised through breeding programs undertaken to develop need-based cultivars. Popular mandarins of different countries are: SATSUMA, PONKAN, DANCY, CLEMENTINE, WILLOW LEAF, KING, CALMONDIN,

CLEOPETRA, TEMPLE, NAGPUR, COORG, DARJEELING/SIKKIM, KHASI, etc.

Satsuma (*Citrus unshiu* Marc.): Small to medium, spreading and drooping tree. Leaves, with large laminae and reduced petioles. Fruit moderately large, oblate to obovate, coarse, with moderately hollow central axis; peel deep orange, sterile ovule, and seedless.

Cultivated in Japan, Spain, Central China, Southern South Africa, Turkey, Korea, Argentina, Uruguay, USA, and several other countries. Common Japanese cultivars are: MIYAGOMA WASE, OKITSU WASE (early-season), and OWARI (mid-season). Fruits are used both fresh and processed ones. Common Chinese cultivars are: XINJIN, GONG CHUAN, NAGA (early-maturing), and WENZHONE also called MIKAN (mid-season type). Japanese OWARI is also cultivated in Spain. Typical Spanish mandarins are CLAUSELLINA, a bud mutation of OWARI (early-maturing), and CLEMENTINE. Characteristics of CLAUSELLINA are similar to those of OWARI except for the time of maturity.

Clementine (*C. clementina* Hort. Ex Tan./*C. reticulata* Blanco): Densely foliated, moderately large tree. Most of its features are common to mandarins but its fruits are usually seedless. There are several biotypes of this cultivar, the common ones of which in Spain are FINA (mid-season), MARISON, OROVAL (early-maturing) and NULE, a late-maturing type. Clementines are also produced in Israel (Plate I: Figure II.1).

Ponkan (*C. reticulata* Blanco): Common Chinese mandarin, an early-maturing type, with low juice acidity and puffy peel. It is also cultivated in Japan, Philippines, and Brazil.

Dancy and Willow leaf (*C. reticulata* Blanco): Common American mandarins. WILLOW LEAF is also called MEDITERRANEAN MANDARIN (*Citrus delicosa* Ten.). DANCY has all the general features of mandarins but 100% nucellar embryony. WILLOW LEAF mandarin trees are densely foliate and compact. Leaves lanceolate with thin petioles. Fruits small to moderate in size, peels tend to become loose and puffy; fruit maturity is mid-season.

Nagpur mandrin: Tree large, vigorous, moderately spreading, spineless, with compact foliage. Fruit subglobose, nonuniform cadmium colored. Surface of fruit smooth; base slightly drawn out with warty glandular furrow; rind thin, soft with slight adherence; pulp vesicles uniformly marigold colored, containing 10–11 separable segments with 6–7 seeds.

Coorg mandarin: Tree vigorous, upright, sparingly spinous, with compact oblate foliage. Fruit globose, bright orange, glossy; base naked or with depressed glandular ribs; thin to medium soft rind with slight adherence, easily peelable; pulp vesicle uniformly colored. Each fruit normally contains 9–11 segments and 14–30 seeds. Cotyledons light green.

Darjeeling/Sikkim mandarin: Tree is evergreen to deciduous, large to medium, in size with dense foliage and occasional thorns. Leaves lanceolate with pointed apex, wingless petioles. Fruit depressed, oval, uniformly bright orange, surface smooth, glossy; base obtuse, occasionally short, naked, slightly ribbed, soft; thin to thick rind not adhering to the pulp. Pulp vesicle uniformly orange, smooth, with 9–11 segments containing 12–20 beaked, slender, slightly plumped seeds. Cotyledons light green.

Khasi mandarin: Medium to tall erect tree, with dense foliage; thorns may be present or absent. Fruit depressed, globose to oblate, orangish-yellow to bright orange, surface smooth, glossy; base even or obtuse, occasionally short, naked, slightly ribbed, soft; rind thin with or without adherence; pulp vesicle contains 9–11 segments, uniformly orange, coarse, with 10–15 seeds.

Kinnow mandarin: Large vigorous erect tree with dense canopy, sparingly thorny. Leaves lanceolate, with round to obtuse base and acuminate apex; petiole short to medium, wingless or narrowly winged. Fruit deep yellowish orange, surface smooth, glossy, shallowly pitted, oblate; base often flattened, slightly depressed, apex much flattened, slightly depressed; thin rind; peel tough and leathery; axis solid to semi-hollow; pulp contains 9–10 segments with 12–24 seeds. Cotyledons yellowish.

In addition to these indigenous naturally occurring diploid types, there are naturally occurring and artificially bred need-based hybrids.

2. Naturally Occurring Hybrids

Common naturally occurring hybrid mandarins are TEMPLE and MURCOTT, also called HONEY. TEMPLE mandarins are natural hybrids between tangerine and sweet orange (tangor). Trees densely foliage with spreading type growth. Leaves lanceolate with thin petioles. Fruit flattened at stylar end, obovate to slightly subglobose, easier to peel than sweet oranges but more difficult than true mandarins, deep orange-colored juice, contain 20–30 seeds and white cotyledons. A seedless selection of TEMPLE in Florida named as SUE LINDA and is cultivated in Florida and South Africa.

MURCOTT is a chance seedling found amongst the natural hybrids of tangor, and first propagated by Murcott in Florida. Trees vigorous and upright. Leaves lanceolate. Fruits produced in terminal clusters, oblate, moderate in size with a hollow central axis, easily peeled, peels not loose or puffy. Both the peel and juice are deep reddish-orange. Cotyledons white. ELLENDALE is a natural tangor, a mid-season cultivar grown in Australia, Argentina, Uruguay, etc.

Fan Qui Rong and Fan QR (2001) discovered a natural hybrid between mandarin and pomelo in China called JINGMENJUYOU. It is a late maturity and dwarf type with high and stable production. Fruits large with longer storage life. Rind of fruit golden-yellow to orangish-red, glossy, easily peeled. Juice is acid-sweet, without bitterness, and with pleasant flavor.

3. Manmade Hybrids

Some of the man-made mandarin oranges are ORLANDO, MINNEOLA, ROBINSON, FALLGLO, LEE, OSCEOLA, NOVA, PAGE ORANGE, SUNBURST, AMBERSWET, FAIRCHILD, etc. ORLANDO and MINNEOLA are hybrids between DANCY tangerine, and DUNCAN grapefruit.

ORLANDO tangelo is a large, vigorous tree with broad cupped leaves and moderate size petioles. Fruits oblate to subglobose, peel bright orange. ORLANDO may be seedless to many seeded, number of seeds varying from 10–20. MINNEOLA is a very large vigorous spreading type tree with large pointed laminate leaves and moderately sized petiole. Fruits are large, obovate with pronounced neck at stem end; peel moderately adherent, finely pebbled; its color, and that of the juice deep reddish-orange. Fruit may be seedless to many seeded, number varies from10–20 (Plate I: Figure II.2)

ROBINSON, LEE, OSCEOLA, and NOVA are hybrids of ORLANDO tangelo, and CLEMENTINE mandarin. PAGE ORANGE is a hybrid of MINNEOLA and CLEMENTINE mandarin. SUNBURST is a hybrid of ROBINSON and OSCEOLA. FALLGLO is a hybrid of BOWER and TEMPLE. AMBERSWET is a hybrid of CLEMENTINE and ORLANDO.

ROBINSON trees are moderately vigorous. Fruit size moderate, shape oblate, color (both internal and external) deep orange, peels easily separable, segments around a hollow open center, seedless or seeded, may contain 20 seeds. FALLGLO is a moderately vigorous, upright and moderately spreading tree. Leaves lanceolate with reduced petioles. Fruits large, oblate with distinct navel; peels smooth with prominent oil glands, easily separable, color of both peel and juice deep orange. AMBERSWET is a moderately vigorous upright tree. Fruits moderately large, convex, tapered at stem end, occasionally produce a small navel, color of juice dark orange; normally seedless, may be seeded, number of seeds may exceed 15. Cotyledons white.

FAIRCHILD is a hybrid of CLEMENTINE, and ORLAND tangelo.

Matsumoto et al. (1999) developed a new hybrid AMAKUSA from a cross between KIYOMI X OKITSU WASE 14, an early ripening mandarin and orange cv Page. They released it as TANGOR NORIN 5 and claimed it to be superior to common tangors. It is seedless with good flavor; rind not very adherent and reddish-orange. Matsumoto et al. (1999 a, b, and c) developed 3 new hybrids: MIKAN NORIN 9, MIKAN NORIN 10 and MIKAN NORIN 11. MIKAN NORIN 9 is a hybrid between ENCORE (KING mandarin X WILLOW LEAF mandarin) and NAKANO 3 PONKAN. It is purportedly a superior mandarin cultivar. MIKAN NORIN 10 is a hybrid between MIHO-WASE (*C. unshui*) and ENCORE. It is claimed to be a superior seedless mandarin resistant to scab and canker. MIKAN NORIN 11 is a hybrid between KIYOMI tangor and NAKANO 3 PONKAN. It too is claimed to be a superior seedless mandarin cultivar resistant to scab and canker.

Roose et al. (2000) developed a seedless late-maturing hybrid from a cross between WILKING and KINEY (WILLOW LEAF/KING/KING/ DUNCAN). They named it GOLD NUGGET. Pio et al. (2000) developed two mandarin hybrids CLEMENTINA CACULA 3 and CLEMENTINA CACULA 4 by crossing CLEMENTINE mandarin with CRAVO tangerine. Fruits of these hybrids are excellent and seedless.

Chen Zhi Min et al. (2001) developed a new Japanese variety derived from a cross between QUINGJIAN Navel orange and ZHONGYE 3 PONGGAN mandarin and named "BUZHIHE. It is a late-maturing type with high fruit-set ratio.

All these man-made hybrids have been developed for potential markets, yield, and resistance to adverse conditions, including pests and diseases (dealt with in Chapters 7 and 8).

Common commercial cultivars of mandarins are listed in the Appendix I.

B. SWEET ORANGE (*CITRUS SINENSIS* OSBECK)

Medium to large tree. Leaf apex bluntly pointed with distinct aroma, petiole narrowly winged. Fruit subglobose to oval, orange colored, tight skinned, with solid central core and smooth orange peel. Number of seeds varies from 2 to 25; seeds oblong, ovoid, planoconvex, generally broad, wedged or pointed at the micropylar end, with oblique ridges surrounding one or two plain areas. Cotyledons whitish.

1. Sweet Oranges of Different Countries

Horticulturally speaking, there are four different types of sweet oranges (a) common or round, (b) navel, (c) pigmented, (blood colored), and (d) acidless.

Round oranges include such cultivars as HAMLIN, VALENTIA, NATAL, PERA, SHAMAUTI, PINEAPPLE, MOSAMBI, and SATHGUDI.

HAMLIN trees are upright and symmetrical. Fruits spherical, small (smaller than other sweet oranges), peel smooth, thin, pale colored, seedless. HAMLIN is usually an early season crop. VALENCIA bears medium size, spherical to oblong fruits, color of the peel and juice deep orange. Fruit seedless. There are different types of VALENCIA, viz., OLINDA, FROST, CAMBELL, MIDNIGHT, DELTA, etc., which differ in quality, fruit shape, peel thickness, and time of maturity. All types of VALENCIA are usually grown as late-season crop. NATAL cultivar of sweet orange is morphologically similar to VALENCIA; trees upright, vigorous and densely foliated; fruits medium size but seeded, containing 5–10 seeds. The NATAL is also a late-season crop. PERA is likewise an upright, vigorous and densely foliated tree. It often shows multiple blooming. Fruit medium size, superior to HAMLIN but inferior to VALENCIA. SHAMOUTI is more upright with broader laminae, ovate fruits with rough thick peel, seedless, and low acidity (Plate I: Figure II.3). It is a mid-season crop a bud mutation of BELADI orange, also known as JAFFA. PINEAPPLE is a mid-season cultivar; tree moderately vigorous, productive, producing deep orange juice but fruits are seedy. Other Common oranges are JINCHENG and XUEGAN of China.

MOSAMBI and SATHGUDI are Indian oranges. Fruits of MOSAMBI are green or yellowish-green, oblate to subglobose or broadly ovoid, aeriolar ring regularly circular, and shallow, base rounded and stalk end set in slight depression, apex broad, even and flat, with a conspicuous but shallow circular furrows; surface smooth with longitudinal furrows; apex with distinct circular ring; rind marigold colored, thin, tight, difficult to peel; pulp apricot yellow, contains 18–25 seeds. Fruits of SATHGUDI are orange colored, spherical, smooth; base and apex evenly rounded; rind thin, semiglossy, finely pitted; pulp uniformly straw colored, segments 10–12, seeds 12–20.

MOSHAMBI is grown in Maharastra and SATHGUDI in Andhra Pradesh.

Navel oranges artificially bred to date include: FISHAR, NEWHALL, LANE LATE, NAVELINA (Plate I: Figure II.4), NAVELATE, SUMMERFIELD, MARSH, BAIANINHA, LENG, PALMER, ROBYN, CARACARA. These oranges are mostly grown in Argentina, Australia, U.S (California), Morocco, Spain, Uruguay, South Africa, Brazil, Venezuela, and several other countries.

Pigmented or Blood Oranges also comprise several selection/artificially bred cultivars. Some important ones are DOUBLE FINA, MALTAISE, SANGUINE, SANGUINELLI, TAROCCO, MORO, etc. Blood oranges are grown mostly in Mediterranean countries.

The fourth group of sweet oranges, called Acidless Oranges, is grown in North Africa, China, Middle East, Brazil, and in some other countries. Common Acidless oranges are: LIMA, SUCCARI, ANLIUCHENG.

Cultivars of oranges developed today are listed in Appendix II.

C. GRAPEFRUIT (*C. GRANDIS* VAR. *RACEMOSA*/*C. PARADISI* MACF)

Tree grows vigorously with spreading growth habit. Leaves unifoliately compound, large, lamina ovate with entire or serrated margin; petiole large, winged, often distinctly cordate. Flower complete, perfect, larger than those found in other *Citrus* species excluding pomelos. Fruits borne in clusters, seedless to seedy. Seeds large, polyembryonic; embryo with creamy to white cotyledons.

There are two general types of grapefruits: white fleshed and red fleshed. Common white-fleshed grapefruits are: DUNCAN (cultivated mostly in the USA), MARSH (cultivated mostly in Cuba, South Africa, and the USA), and OROBLANCO (SWEETIE) (mostly cultivated in the USA and Israel). Common red-fleshed or pigmented grapefruits are: FLAME (cultivated mostly in the USA) (Plate I: Figure II.5), REDBLUSH or RUBY RED (cultivated mostly in South Africa, Cuba, and the USA); RIO RED (cultivated mostly in the USA), STAR RUBY (mostly cultivated in the USA and South Africa). Other pigmented grapefruits are HENNINGER, BUR GUNDY, FOSTER, HUDSON, THOMPSON, HENDERSON, RAY RUBY.

Luth and Moore (1999) developed a transgenic grapefruit (*Agrobacterium tumefaciens* mediated) and opened a new chapter for qualitative and quantitative improvement of citrus in future.

Common commercial cultivars of grapefruits are listed in Appendix III.

Saunt (1990) was of the opinion that all pink-fleshed cultivars originated from white-fleshed grapefruits. According to him, seedy white-fleshed grapefruits probably were brought from Barbados to Florida (USA) in 1823 and subsequently all other cultivars arose by limb sport or selection. He postulated the pedigree of the different cultivars as follows:

"Original seedy white-fleshed grapefruit > Duncan: seeded white cultivars originated as seedling in 1832 (named after the discoverer); White Marsh: chance seedless seedling originated in 1880 (named after the propagator); Walters: seedy white (originated in 1887).

White Marsh > Thompson: seedless, light pink fleshed, originated as limb sport in 1913 (named after the propagator).

Thompson > Red blush: seedless, red-fleshed, originated as a limb sport of Thompson in 1931; Ruby Red: seedless, red, originated as a limb sport in 1929.

Walters > Foster: pink, seedy, originated in 1907 as a limb sport.

Foster > Hudson: red, seedy, originated in 1930 as bud sport

Hudson > Star Ruby: deep red, seedless, through irradiation and selection in 1959

Ruby Red > Rio Red: deep red, seedless; originated in 1984; Red blush; Ray Ruby: deep red, seedless, originated in 1977; Henderson: deep red, seedless, originated in 1973.

Henderson > Flame: deep red, seedless, originated in 1987"

D. POMELO (SHADDOCK) (*CITRUS GRANDIS* (L.) OSB.)

Tree grows vigorously and bears the largest leaves, flowers and fruits among all the *Citrus* species. Leaf lamina ovate with serrated margin, petiole cordate with distinct wing separated from lamina; young leaves and midrib pubescent. Fruit ovoid to pyriform, stylar end flattened. Peel extremely thick, easy to peel; juice sacs pronounced, rubbery in texture. Juice sweet with mild flavor. Seeds large, plump. Progeny is only zygotic. Fruits are available in various shapes, sizes, flesh colors (white, pink red, green, yellow) with excellent storage capability.

Pomelos are mostly cultivated in China, Japan, Indonesia, and several Southeast Asian countries. There are numerous cultivars of this species; the important ones are: BANPEIYU, CHANDLER, GOLIATH, KAO PAN, KAO PHUANG, MATOYU, NAKAN, PANDAN WANGI, SHATIANU, THONG DEE.

Long and Long (2000) discovered LONGYAN HONGYOU a sport of pomelo (*C. maxima*) cv *Guanximiu.* Fruits are large and have a smooth orange-yellow rind. The flesh is purplish red, tender, very juicy with a pleasant acid-sweet flavor and top eating quality. 100 ml of the juice contains 43.03 mg ascorbic acid.

E. ACID LIME (*C. AURANTIFOLIA* SWINGLE)

Tree grows vigorously with upright to spreading growth, thorny. Lamina large elliptic to ovate with crenate margin, obtuse apex, prominent veins and distinctly articulated narrowly winged or reduced petioles. Flowers perfect, complete; petals white or purple colored, may be extrorse, curving downwards from the central axis. Fruits yellow, oblong to round, smooth with slightly nippled rounded apex, rounded base. Rind thin, papery with strong or low adherence; containing 9–11 segments, light greenish-yellow, coarse pulp with 9–15 seeds.

There are two types of Acid limes, TAHITI (lamina larger, petals purple, fruit similar to lemon, spheroid, rarely produce seed) and WEST INDIAN LIME, also known as MEXICAN, KEY, LIMA, KAGHZI, GALEGO, LIMUNBELADI LIME, etc. (lamina smaller, ovate to round, distinctly serrated around the apical margin, flower smaller, white, fruit smaller, spherical, moderately seeded, distinctive aroma and flavor, juice highly acidic).

TAHITI LIMES are grown in USA, Mexico, Australia, Brazil, and several other countries. WEST INDIAN LIMES are grown in Brazil, Mexico, Egypt, India, and several other countries. Limes grown in various countries are given in Appendix IV.

F. LEMON (*C. LIMON* (L.) BURN. F.)

Lemon is a hybrid and closely related to Citron (Scora, 1988). It is an extremely vigorous tree with upright and unwieldy growth habit, and thorny nature that varies with cultivars, growing conditions and age. Leaf morphology variable with tree vigor; lamina large, ovate with pronounced serrations along the apical margins; newly developing leaves purple but mature ones green, petioles reduced or nonexistent. Flowers complete, perfect, purplish-white, borne in clusters. Fruit spherical to prelate spheroid with apical nipple at stylar end; may be seeded or seedless. Seed-coat smooth with pointed mycropylar end. Juice highly acidic.

There are several cultivars of lemons. Common ones are: EUREKA, FEMINELLO, FINO (also known as PRIMOFIORI, MESERO, BLANCO, etc.), INTERDONATO, LISBON, MONACHELLO, VERNA (also known as BERNA).

In Northeastern India, several types of lemons are found, of which the common ones are HILL LEMON (*C. pseudlimon* Tanaka), NEPALI OBLONG, and ASSAM LEMON (*C. assamensis* Dutta).

Eureka Lemon: Tree medium size spreading, thornless. Leaves dark green, margin crenate, apex blunt, pointed or rounded. Fruit oblong or elliptic to ovate with gradual rounded base which may be necked; apex may be abruptly rounded or shouldered, yellow, surface slightly rugose, pitted, pulp in 9–10 segments, grained, containing 0–6 seeds.

Lisbon Lemon: Tree grows vigorously growing with dense foliage and upright thorny branches. Fruit smooth with drawn out nipple, ellipsoid or oblong in shape, and tapering base; pulp greenish yellow, 7–10 segments containing 0–10 seeds.

Hill Lemon (*Citrus pseudolimon* Tanaka): Tree hardy, vigorous, fairly tall, upright and spreading with dense foliage and light green thorns. Leaf elliptic, ovate with crenate margin, obtuse acute apex; petioles marginally winged and long. Fruit oblong-ovate with nippled apex and rounded or slightly nippled base, rind thick, pulp light-yellow, coarse, 8–10 segments containing 28–60 seeds.

Nepali Oblong Lemon: Tree hardy, vigorous, spreading and drooping with dense foliage, and sparse light green thorns. Leaf elliptical with dark green upper surface, serrated margin, acute apex and narrowly winged petiole. Blooms throughout the year. Fruit oblong to ovate with nippled apex and rounded base, lemon-yellow; pulp with 11–13 segments, light yellow, fine, containing a few or no seeds.

Assam Lemon (*Citrus assamensis* Dutta) Tree semi-wild with glossy and aromatic (ginger or Eucalyptus-like aroma); fruit spherical.

Common commercial cultivars of lemons know to date are listed in Appendix V.

Varietal Improvement

Cultivar improvement programs are being carried out in different countries depending on the market demand. Fruits are mostly consumed fresh or used to meet the demand of the processing industries. Fresh fruit industry promotes organoleptic qualities (aroma, taste, and acidity) and horticultural qualities (easy to peel, seedless, color, rind appearance, etc.). Fruits for processing industry must have good percentage of juice and sugar content. Agronomically, extending the production periods and disease resistance have top priorities. Recent development in biotechnological methods such as somatic embryogenesis, haploidization, somatic hybridization, genetic transformation, including several diagnostic techniques such as biochemical and molecular markers, flow cytometry, embryo rescue, etc. have opened new vistas for breeding market-specific new cultivars.

REFERENCES

Chen ZhiMin, Quian Jiebing, Yang JiaDong, Chen ZM, Qian JB, and Yang JD (2001). The characteristics of Buzhihe citrus variety and its cultural techniques. *South China Fruits* 30 (4): 6–7.

Davies FS and Albrigo LC (1994). Citrus. CAB International. Willingford, UK.

Fan Qui Rong and Fan QR (2001). The characteristics of Jinglongjuyou mandarin variety. *South China Fruits* 30: 4.

Long YanYou and Long YY (2000). "Longyan Hongyou" a sport of Guanximiu pummelo variety. *South China Fruits* 29 (2): 6.

Luth D and Moore G (1999). Transgenic-grapefruit plants obtained by *Agrobacterium tumefaciens*-mediated transformation. *Plant Cell Tissue* and *Organ Culture* 55(3): 219–222.

Matsumoto R, Okudai N, Yamamoto N, Yamada Y, Asada K, Oiyama I, Ikemiya H, Takahara T et al. (1999). A new citrus cultivar: 'Amakusa'. *Bull. Fruit Tree Science* 33: 37–46.

Matsumoto R, Okudai N, Yamamoto M, Yamada Y, Takahara T, Oiyama I, Ishiuchi D et al. (1999a). A new citrus cultivar: 'Hareyaka'. *Bull. Fruit Tree Science* 33: 47–56.

Matsumoto R, Okudai N, Yamamoto M, Takahara T, Oiyama I, Ishiuchi D et al. (1999b). A new citrus cultivar: 'Miho-core'. *Bull. Fruit Tree Science* 33: 57–66.

Matsumoto R, Okudai N, Yamamoto N, Takahara T, Yamada Y, Oiyama I, Ishiuchi D et al. (1999c). A new citrus cultivar: 'Youkou'. *Bull. Fruit Tree Science* 33: 67–76.

Pio EM, Minamik F, Figueiredo Jo de, Pompeu Jr., and de Figuiredo JO (2000). Characterization and evaluation of two new hybrid varieties of 'Clementine' mandarin. *Laranja* 21 (1): 149–159.

Roose ML, Williams TE, Cameron JW, and Soost RK (2000). 'Gold Nugget' mandarin: a seedless late maturing hybrid. *HortSci.* 36 (6): 1176–1178.

Saunt J (2000). Citrus varieties of the world: An illustrated guide. Sinclair International, UK.

Timmer LW and Duncan LW (Eds.) (1999). Citrus Health Management. American Phytopathol. Soc., Citrus Research and Education Center, Lake Alfred, University of Florida, Fla, USA.

3

Rootstocks and Their Uses

Most of the *Citrus* species are perennial trees initially found in forest habitats that were later domesticated and improved for commercial uses. Many of these trees can survive for more than 100 years but with age, production gradually becomes uneconomical. Juvenility of these domesticated species is high and they start fruiting after six years. Their specific productivity usually appears after ten years. To remove juvenility and to achieve early uniform fruiting, grafting/budding was introduced in citriculture during the nineteenth century. Historically, Sour orange was the standard rootstock in citrus-growing regions because of its tolerance to *Phytophthora* foot rot. Gradually rootstock became an important component in advanced citriculture to resolve field problems and to improve commercial qualities. Japan had the problem of cold-hardiness for seedless Satsuma oranges but eventually found the use of Trifoliate orange *(Poncirus trifoliata)* resolved this problem. Similarly, Rough lemon was introduced in the USA as rootstock for growing citrus in sandy soil and remained predominant till the blight disease started to take a heavy toll of the crop. Standard rootstock sour orange also got a setback because of its susceptibility in combination with sweet orange to the Citrus tristeza virus. Thus rootstock research began to gather momentum and has been primary focus in citriculture since the 1980s.

A. SIGNIFICANCE OF ROOTSTOCKS

Rootstock selection is a major consideration in every citrus growing operation. The rootstock chosen provides the root system of the budded/grafted tree that remains responsible for anchorage of the plant, absorption of nutrients, water, and also storage of carbohydrates produced in leaves, synthesis of some growth regulators, and source of tolerance/resistance to diseases and pests. More than 20 horticultural characters are normally influenced by rootstocks, in particular size, vigor, depth of rooting, frost tolerance, adaptation to certain soil conditions such as high salinity or pH, excess water, resistance or tolerance to pests, diseases, fruit characters such as yield, size, texture, internal quality and maturity date. Selection of rootstocks depends on critical limiting factors to production existing in the growing region, in particular climate, soil, predominant pests, diseases, and the mother cultivars. For example, sour orange is not to be used as a rootstock where the tristeza virus prevails, although it provides excellent vigor to the scion. Similarly, rough lemon is normally avoided in regions affected by frost. Trifoliate orange although providing protection from *Phytophthora* performs poorly in soil of high pH; it is well adapted to cool growing conditions and acid soil but growth remains moderate. Carrizo provides protection from burrowing nematodes. Details of available rootstocks, their properties, and scionic combinations are dealt with separately. All rootstocks are not equally adaptable to all scionic cultivars. Mandarins and their hybrids do well with Cleopetra mandarin but sweet orange and grapefruit cultivars do not. Cleopetra, on the other hand, is quite tolerant to tristeza virus, and moderately tolerant to blight. Rootstocks that provide high vigor generally give higher yield but the quality may be poor.

Rootstock Root System and Its Implications

The success of a rootstock depends to a large extent upon its root system. As the primary root emerges from the seed, root hairs often appear. Citrus roots are normally taprooted. During germination the radicle generally appears first and rapidly grows downward, forming a well-defined taproot, if left undisturbed. It is a common practice, however, to sever the taproot during digging the nursery trees and it may lose the strong identity evident in the seedling stage. The range of the development of the root system depends both upon the rootstock species and the soil type. The natural tendency of citrus seems to be formation of a bimorphic distribution characterized by a network of numerous, relatively shallow lateral roots that provide the supporting framework for a dense mat of fibrous roots and smaller laterals. The second grouping is more or less vertically oriented. Laterals may develop to lengths 2 or 3 times of the canopy diameter while remaining uniform in thickness throughout their length. The smallest roots have been labeled feeder or fibrous roots and rootlets. This normal root system often becomes modified by various factors. In deep sandy soil, vigorous rootstocks have extensive systems with an abundance of fibrous roots; less vigorous rootstocks have shallower root systems with the major portion of their fibrous roots concentrated near the surface; these differences again are not constant and vary with soil types. Root growth is vigorous in sands except when impeded by layers of high organic or clay content. In soil with a high water table, the root system tends to "pancake" with nearly 75% of the fibrous roots located near the surface. In other soil types such as loam, fewer roots are required to support a tree of comparable size growing in sand.

B. COMMON ROOTSTOCKS

After introducing the budding/grafting system of propagation, suitable natural *Citrus* species and their relatives were used as rootstocks. Subsequent extensive research led to the development of selections, mutants, hybrids, etc. that could be used as rootstocks (Castle, 1987; Davis and Albrigo, 1994; Aubert and Vullin, 1998; Castle and Gmitter, 1999). Most commonly used rootstocks are Sour orange, Rough lemon, Ranpur lime, Trifoliate orange, Citranges, and Cleopetra mandarins.

Sour Orange (*Citrus aurantium* Linn.)

Medium-size tree, with dense, compact, rounded canopy; thorny. Foliage evergreen, articulated, alternate, ovate, pointed, distinctly scented; petiole broadly winged. Flowers small, in axillary cymes, white, scented with conspicuous oil glands. Fruits, medium size, compressed, orange colored; rind strongly aromatic, thick, bitter, easily separable; juice acidic with sour taste. Seeds few to numerous, flattened, wedged at micropylar end, marked with ridged lines, measuring 11–14 mm in length and 6–10 mm in breadth, chalazal spot creamy red, surface rough, veined, wrinkled; cotyledons light green.

Sour orange possesses tolerance to *Phytophthora,* frost, blight, exocortis, and xyloporosis but susceptible to mal secco, tristeza, citrus nematode and burrowing nematode. It is claimed to be a symptomless carrier of tatter leaf and exocortis. It is not adaptable to waterlogging and heavy soils.

Sour orange exhibits good productivity, yield, and fruit quality. It is compatible with all citrus varieties except kumquats and certain mandarins of the Satsuma type. Fruits from budded plants show rind thickness, high juice content, total soluble solids, and average acidity. The plants show acceptable tolerance to chloride and boron and satisfactory tolerance to calcium and poor drainage. Their root distribution is deep. Common Sour orange and hybrids used as rootstocks are given in Appendix VI.

Rough Lemon (*Citrus jambhiri* Lush)

Medium to large tree with spreading growth habit; with short, stiff sparse spines and pale new growths. Leaves small to moderately large, light green, lanceolate and acuminate or round with subserrate margin and medium petiole. Flower small, faintly colored, with numerous stamens. Fruit light yellow to lemon brown, oval, pointed at both ends with fleshy apical papilla and strongly developed apical cavity surrounding the papilla, with rind fairly loose with rough and irregular surface. Core of fruit solid opens at maturity, flesh yellow. Seeds few to numerous, small with white embryo; ovoid or slightly tapering, measuring 10–11 mm in length and 5–6 mm in breadth; surface smooth, veined, cream colored; chalazal spot brownish red (Plate II: Figure III.1).

Rough lemon is suitable for deep sandy soil; its root system is extensive, sometimes reaching a depth of 4–6 m in deep sands.

Trees are very drought tolerant, moderately tolerant to high salinity and well adapted to a wide range of soil pH, with high tolerance to loam but susceptible to clay, acceptable tolerance to boron, high tolerance to calcium but susceptible to poor drainage. Sweet orange, grapefruit, lemon, and mandarin are very productive on this rootstock. Total solid production/tree/ha is very high. Though the quality does not satisfy the parameters for marketing as fresh fruit. It is in great demand for processing industries. Fruit peels tend to be thick and puffy and hold poorly on the tree. Trees show unsatisfactory cold hardiness and longevity. Plants show 90–100% nucellar embryony.

Plants susceptible to fruit damage, tolerant to tristeza, exocortis, psorosis, xyloporosis, and cachexia; severe strain of tristeza can produce stem pitting; highly susceptible to *Phytophthora,* citrus and burrowing nematodes, and blight. Grapefruit when grafted on Rough lemon shows low chilling injury (Renaldo, 1999).

Common Rough lemons and hybrids commonly used as rootstocks are given in Appendix VII.

Trifoliate Orange (*Poncirus trifoliata*)

Small deciduous tree, densely branched, with upright habit and stout, stiff and sharp thorns. Foliage trifoliate, elliptical, crenate; borne singly or in tuffs. Flowers single or in pairs, axillary, almost sessile, normally appear before the foliage; petals white, large and ovate. Fruit light orange, rough, densely covered with short hairs, glandular with aromatic oils; rind thin, juice content high, total soluble solids very high, acidity normal; seeds brownish at the broad end.

The plant has satisfactory tolerance to *Phytophthora,* citrus nematodes, and woody galls; it is symptomless to psorosis, cristacortis, concave gum, cachexia, xyloporosis, and woody gall; slightly tolerant to tristeza stem pitting; susceptible to burrowing nematodes, exocortis, tatter leaf, and blight. Trees are susceptible to sand, tolerant to heavy soil, loam and clay, and moderate waterlogging; highly sensitive to calcium and chlorides; fruit quality satisfactory with satisfactory cold hardiness, longevity, and compatibility for all commercial *Citrus* species, but fruit size and vigor nonacceptable.

Poncirus flying dragon, a clonal selection of *P. trifoliata,* is a fully dwarfing rootstock forming compact or stunting canopy. Trifoliate oranges and hybrids commonly used as rootstocks are given in Appendix VIII.

Rangpur Lime (*Citrus limon* Usbek)

Rangpur lime is probably a natural hybrid between true lime and mandarin. It is a medium size tree with spreading habit and round top; twigs slender, thorns short, sparse. Foliage dark green, elliptical, less pointed, margin with shallow serration. Flowers small, purple tinged. Fruit small to medium size, in clusters, reddish deep orange, globose or round, nippled; rind thin, moderately loose and loosely adherent; juicy, juice strongly acidic. Seeds small, oval, pointed at micropylar end, surface smooth, veined, cream colored with yellow-lime cotyledon and brown chalazal spot.

The notable good characteristics of *C. limon* are vigorous growth, high yield, early fruiting, high fruit caliber, good color, high juice content, average soluble solids, and average acidity. The plant is susceptible to *Phytophthora*, citrus and burrowing nematodes, blight, cachexia, exocortis, and woody gall; mildly tolerant to xyloporosis, concave gum, and tristeza stem pitting; symptomless carrier of psorosis, cristacortis, tristeza, sensu stricto, and tatter leaf.

Further, Rangpur lime has high tolerance to chlorides, moderate tolerance to boron, calcium, sand, loam, and clay. Its root system is deep. It is cold hardy, with good longevity, acceptable fruit size, quality, and compatibility.

Volkamer Lemon (*Citrus volkameriana* Tan & Pasq.)

This lemon hybrid is a large, vigorous tree that like Rough lemon yields a large quantity of moderate to poor quality fruits. Fruiting early with good fruit caliber; rind thick; average fruit color, juice content, total soluble solids and acidity.

It is susceptible to nematodes, blight and woody galls; symptomless carrier of tristeza, sensu stricto, tatter leaf, exocortis, psorosis, cristacortis; slightly tolerant to cachexia, xyloporosis and *Phytophthora*, with low cold tolerance. The plant has good tolerance to chlorides and adapts well to light soil and also adaptable to various other soil conditions. Fruit quality of mandarins and oranges on this stock is poor.

Alemow (*Citrus macrophylla* Wester)

A hybrid between *Citrus celebica* and *Citrus grandis*, it is morphologically and genetically similar to lemons and limes. Cultivars budded on this plant produce large, vigorous, and high yielding trees except Eureka lemon, and show early fruiting; average fruit caliber, thick rind, average juice content, and total soluble solids with low acidity. The plants have low cold tolerance and are susceptible to nematodes, blight, tristeza sensu stricto, cachexia and xyloporosis. The plant is a symptomless carrier of exocortis, psorosis and cristacortis; it is tolerant to *Phytophthora*. It is excellent for lemons and limes. It adapts well to cool, dry climate; sandy, calcareous soil, high pH, and has a deep, dense root system and it is drought resistant and frost insensitive. The fruits of sweet oranges and grapefruits on it are large, puffy, and the quality moderate to poor. It can tolerate a high level of boron, chlorides, and calcium, and is adaptable to wide range of soil conditions but accumulates manganese.

Palestine Sweet Lime (*Citrus limettoides* Tan.)

The Palestine is not a true lime but a hybrid and is not widely used as a rootstock. Plants have high vigor and productivity with large fruits. This lime is susceptible to *Phytophthora,* citrus and burrowing nematodes, exocortis, tristeza, cachexia, chlorides and boron but tolerant to calcium; its performance in sand and loam is satisfactory but not in clay. Its fruit is satisfactory but not the quality. It is a large and vigorous tree with a deep root system and satisfactory cold hardiness.

Sweet Oranges (*Citrus sinensis* (L.) Osbeck)

Sweet oranges are mostly used as scions. Botanical and horticultural characters of the species have been described Chapter 2. When these plants are used as rootstocks, the combinations show moderate vigor with good crops and quality fruits. In addition, they also show blight tolerance and can develop deep widespread root systems. Further, they are tolerant to tristeza, exocortis, and xyloporosis but susceptible to *Phyotphthora*, citrus and burrowing nematodes, flood and drought. "Caipira DA" sweet orange is regarded as a good rootstock (Koller et al., 2000).

Mandarins (*Citrus reticulatus* Blanco)

Botanical and horticultural characters of mandarins have been described in Chapter 2. Several cultivars of these plants have found use as good rootstocks. The most popular one in tropical countries is Cleopetra mandarin. These plants are tolerant to *Phytophthora*, tristeza, exocortis, and psorosis but susceptible to citrus and burrowing nematodes; there is, however, a controversial report on its susceptibility to cachexia. It has tolerance to boron, calcium and chlorides; it performs well in loam and clay but its performance is unsatisfactory in sandy soil. Oranges and grapefruits grafted/budded on Cleopetra show acceptable vigor, tree size, root distribution, cold hardiness, and fruit quality but unsatisfactory fruit size. Another mandarin used as rootstock is Fuzhu mandarin. It is tolerant to *Phytophthora*, cachexia, xyloporosis, concave gum and nematodes, and a symptomless carrier of tristeza, sensu stricto, tatter leaf, psorosis, and cristacortis. Plants are vigorous with high yield, early fruiting, and cold hardiness. They show good fruit caliber, thin rind, good fruit color, high juice content, total soluble solids and normal acidity.

Other mandarin rootstocks are Nasnaran (*Citrus amblycarpa),* Shekwasha mandarin (*Citrus depressa*), and Sun Chu Sha. Nasnaran gives higher yield than Cleopetra; its disadvantage is that it is very thorny at the seedling stage. Shekwasha is mostly used in a breeding program because of its high tolerance to iron chlorosis normally induced at high pH. Sun Chu Sha was initially claimed as tolerant to blight, tristeza and *Phytophthora* and performing well in magnesium deficient soil, and highly tolerant to iron chlorosis; it was later found to be susceptible to *Phytophthora (*Timmer and Duncan 1999).

Mexican Lime

Botanical and horticultural characteristics of this cultivar have been described Chapter 2. It is normally used as rootstock for oranges and grapefruits. The plant has highly satisfactory vigor, satisfactory tree size and compatibility but unsatisfactory cold hardiness. It is tolerant to boron, chlorides and calcium but susceptible to *Phytophthora,* tristeza, citrus and burrowing nematodes.

Citranges (Hybrid of Trifoliate orange and Sweet orange)

Two types of citranges are normally used as rootstocks, Troyer Citrange and Carrizo Citrange. Washington navel sweet orange is normally used to breed commercial citranges.

Troyer Citrange: Named after Mr. A. Troye, a nurseryman of the USA who first made a tree derived from the original hybrid in 1931, this hybrid is suitable for humid climate. The plant has average vigor, good yield, early fruiting, and cold tolerance; it has average fruit caliber, good fruit color, thin rind, high juice content, total soluble solids and normal acidity, and acceptable fruit size and quality. It has moderate root distribution and tolerance to boron but is susceptible to chloride and calcium. It has tolerance to *Phytophthora,* tristeza, and nematodes but is susceptible to blight, tatter leaf, and exocortis. The plant is a symptomless carrier of sensu stricto, psorosis, cristacortis, cachexia, and xyloporosis; it has also been claimed to be a symptomless carrier of tristeza.

Carrizo Citrange: Carrizo Citrange hybrid seedlings were first raised in a nursery located near the Carrizo springs in Texas, USA and named according to its place of origin in 1983. The plant has average vigor, high yield, and average cold tolerance; it has average fruit caliber, good fruit color, thin rind, high juice content, total soluble solids, and normal acidity;

fruit size and quality are satisfactory. Root distribution is moderate. The plant is tolerant to boron but susceptible to chloride and calcium. It is highly tolerant to *Phytophthora*, nematodes, and tristeza stem pitting; susceptible to blight, exocortis, and a symptomless carrier of psorosis, cristacortis, cachexia, and xyloporosis. It is claimed to be a symptomless carrier of tristeza and sensu stricto.

Citrumelos: Citrumellos are hybrids of *Citrus paradisi* and *Poncirus trifoliata*.

The common citrumelo, widely used as rootstock, is Swingle citrumelo or Citrumelo 4475. This plant has average vigor, high yield, and average cold tolerance; it has average fruit caliber, thin rind, high juice content, and average total soluble solids with normal acidity. It is tolerant to *Phytophthora*, nematodes, cachexia, xyloporosis, blight, and tristeza stem pitting; and is a symptomless carrier of psorosis and cristacortis. It is also claimed to be symptomless carrier of tristeza, and sensu stricto but is susceptible to tatter leaf and exocortis. It is further claimed to be resistant to exocortis (Castle and Gmitter, 1999).

Several new rootstocks have recently been developed. Most successful among them is "HRS 812" for both oranges and mandarins (Bowman et al., 2000).

The adaptability of common conventional rootstocks to different biotic and abiotic stresses is given in Appendixes IX and X.

Use of Rootstocks in Different Countries

Use of rootstocks mostly depends upon the properties of soil, its nutritional status, climate, and biotic and abiotic stresses of the region. Accordingly, various countries select rootstock species for their preferred citrus crops. Witscher and Crane (2000) recommended the use of trifoliate and its hybrids, citranges, Swingle citrumelo, Rangpur lime and Cleopetra mandarins in tropical countries. Zekuri (2000) found Swingle citrumelo be the best-suited rootstock for high density planting in Florida.

Different types of commonly used scionic species were described in the previous chapter. The rootstock scionic combinations cultivated in various countries are given in Appendix XI.

Varietal Improvement

As in the case of current varietal improvement programs for commercial cultivars, improvement of rootstocks, following similar biotechnological methods, is also being conducted in different countries. The driving force for these programs is not commercially oriented. These programs are need-based, particularly for adaptation to different types of soil and climate and for disease resistance. Compatibility to preferred scionic cultivars is also a matter of concern.

REFERENCES

Aubert B and Vullin G (1998). Citrus Nursery and Planting Techniques. CIRAD, France, pp. 46–52.

Bitters WP and Parker ER (1951). Rootstocks investigations. *California Citrograph* 36: 313, 329–330.

Bowman KD, Wutscher HK, Kaplan OT, and Chaparro JX (2000). A new hybrid citrus rootstock for Florida: US 852 *Proc. Florida State Hortic. Soc.* 112: 54–55.

Castle WS (1987). Citrus rootstocks. *In*: Rootstocks of Fruit Crops (eds. RC Rom and RS Carlson). John Wiley and Sons, New York, NY, pp. 361–399.

Castle WS and Gmitter FG (1999). Rootstocks and scion selection. *In*: Citrus Health Management (eds. LW Timmer and LW Duncan). APS, Citrus Research and Education Center, Lake Alfred, Fla, pp. 27–34.

Davis FS and Albrigo LG (1994). Citrus. CAB International, Wallingford, UK.

Forner JB, Pina JA, Aparicio M, Sala J, Alcaide A, and Giner J (1981). Recent Status of Citrus Rootstocks in Spain. *Proc. Int. Soc. Citriculture*, Int. Citrus Cong. Tokyo, Japan, vol. 1, pp. 106–198.

Jacquemond C and de Rocca Serra D (1992). Citrus Rootstock Selection in Corsica for 25 years. *Proc. Int. Soc. Citriculture*. VII Int. Citrus Cong. Acireale, Italy, vol. 1, pp. 246–255.

Koller OL, Soprano E, Costa ACE de, and de Costa ACZ (2000). Evaluation of rootstocks for 'Hamlin' orange in Santa Catarina, Brazil. *Revista Ceres* 47: 271, 325–336.

Newcomb DA (1978). Selections of rootstocks for salinity and disease resistance. *Proc. Int. Soc. Citriculture*. University of Sidney, Australia, pp. 117–120.

Pasos OS and de Cunha Sobrinho AP (1981). Citrus rootstocks in Brazil. *Proc. Int. Soc. Citriculture*. Int. Citrus Cong. Tokyo, Japan, vol. 1, pp. 102–105.

Reynaldo IM (1999).The influence of the rootstock on the post-harvest behavior of Ruby Red grapefruit. *Cultivos Tropicales* 20 (2): 37–40.

Russo F and Reforgiato Recupiro G (1984). Recent results of some Citrus rootstock experiments in Italy. *Proc. Int. Soc. Citriculture*. Int. Citrus Cong., Sao Paulo, Brazil, vol. 1, pp. 42–44.

Shaked A and Ashkanazy S (1984). Swingle Citrumelo as a new citrus rootstock in lsrael. *Proc. Int. Soc. Citriculture*, Int. Citrus Cong., Sao Paulo, Brazil, vol. 1. pp. 48–49.

Simon A, Jimenez R and del Valle N (1992). Rootstocks for grapefruits (*Citrus paradisi* Macf.) in Cuba. *Proc. Int. Soc. Citriculture*, VII Int. Citrus Cong., Acireale, Italy, vol. 1., pp. 262–264.

Witscher HK and Crane JH (2000). Appropriate use of scion/rootstock combination for citrus in the tropics. *Proc. Int. Soc. Tropical Horticulture* 42: 166–172.

Zekuri M (2000). Citrus rootstocks affect scion nutrition, fruit quality, growth and economical return. *Fruits* 55 (4): 231–239.

4

Polyembryony and Detection of Embryos

Most species of *Citrus* are apomictic and polyembryonic. This implies that when a seed germinates, it produces more than one seedling. Seedlings that grow from *nucellar embryos* appearing in the maternal tissues of the ovary are called *nucellar embryonic seedlings*. Seedlings that grow from *zygotic embryos* (products of the union between male and female gametes) are called *zygotic embryonic seedlings*. The nucellar seedling is true to type to the mother tree but the zygotic seedling inherits characteristics of both the parental trees. These seedlings are also called *gametic seedlings*.

A. POLYEMBRYONY IN VARIOUS *CITRUS* SPECIES

Strasburger in 1878 first observed the regular occurrence of polyembryony in certain species of *Citrus* in nature. Subsequently, Webber in 1920 observed the utility of nucellar seedlings in the propagation of nursery trees as they are true to the mother type. Although polyembryony is a common phenomenon in citrus seeds not all the species exhibit polyembryony to the same extent between and also within them (Table 4.1). Some species of *Citrus* are, however, monoembryonic. These are *Citron, Pomelo, Clementine mandarin,* and *Bergamot*. Variation in the rate of polyembryony within a species did not draw the attention of many citriculturists. Comprehensive data are available in the study conducted by Pio et al. (1984). They worked out the polyembryony of 26 cultivars of Trifoliate Orange and found wide variation in the extent of polyembryony both by directly counting the embryos in seeds and counting the seedlings after their emergence (Table 4.2).

These data indicate that the extent of polyembryony of species/clones may differ in different agroclimatic and climatic regions.

Regional variation of the extent of polyembryony is apparent from the Tables 4.1 and 4.2. These tables show that the number of embryos per seed usually does not differ widely. Pio et al. (1984) recorded the average number of embryos per seed of 26 clones of Trifoliate Orange as 1.58–3.26

Table 4.1: Extent of polyembryony (%) in different *Citrus* species as reported by different authors

Species	1	2	3	4	5
Rough lemon	x	91	70	100	94
Trifoliate orange	72	71	35	x	58
(Flying dragon)					
Poncirus &					75–90
Citrange					
Lemon	10–95	x	x	x	33
(*Eureka*)					
(*Femminello*)					69
Sweet Orange	40–45	x	50	40–95	98
(*Washington*)					
(*Valencia*)					79–85
Mandarin	10–100	x	42	10–100	
(*King*)					11–21
(*Dancy*)					95
(*Cleopetra*)					90
Grapefruit	60–95	x	nil	60–95	96
(*Marsh*)					
Acid lime/	x	x	62	78	
Rangpur lime					
Sour Orange	x	x	49	75–85	x
Tangelo	x	x	x	x	
(*Orlando*)					83

1 = Webber (1932), 2 = Randhawa and Bajaj (1958), 3 = Pradhan (1993), 4 = Rajput and Haribabu (1995), 5 = Aubert and Vullin (1998)

on direct count and 1.04–1.30 on counts after emergence of seedlings in seedbeds. Pradhan (1994) in her studies with different species of *Citrus* observed 1–5 embryos per seed after their germination in vitro.

B. DETECTION OF NUCELLAR AND ZYGOTIC SEEDLINGS

Ebrahimi (1984) did a comprehensive study on polyembryonic seeds of nine varieties of citrus and compared their morphological characters. He continued his studies with seedlings of equal height (some seeds produced two to three seedlings of equal height and one shorter; some seeds produced two to three seedlings of equal height and one taller; whereas some seeds produced only two to three seedlings of equal height). Uniform seedlings of equal size and height were grown and compared with the mother plants at their bearing stage. He found that trees from equal-size seedlings were similar to their mother trees and were all *nucellar*. This study formed the basis for detecting *nucellar* seedlings in seedbeds by their uniformity in size and is still practiced by nurserymen today.

Detection of *nucellar* seedlings in seedbeds following the above method is a time-consuming process. Attempts have been made to overcome this

Table 4.2: Polyembryony in 26 cultivars of Trifoliate Orange (Pio et al., 1984)

Clones/Cultivars	Direct observation		After emergence in seedbed	
	Av. No. of embryos Per seed	Percentage of polyembryony	Av. No. of embryos per seed	Percentage of polyembryony
E.E.Limeira	2.06	76	1.26	26
Rich 22–2	1.58	52	1.06	6
Davis B	1.78	64	1.04	4
Rubidoux	1.98	68	1.08	8
Kryder 8–5	1.68	62	1.08	8
Rich 21–3	1.78	56	1.10	10
Taylor	1.88	74	1.16	16
Davis A	1.86	70	1.86	10
English-large	1.84	62	1.18	18
Christian	3.00	100	1.14	16
Rich 16–6	2.18	80	1.06	6
Argentina	2.08	80	1.30	26
Towne F	1.64	62	1.10	10
Jacobsen	1.96	70	1.04	4
English-small	1.88	72	1.24	24
S. Joaquim Valley	2.64	88	1.40	36
Kryder 28–3	1.98	52	1.06	4
Benecke	2.34	98	1.12	12
Rich 5–2	3.26	100	1.10	10
Tucuman	3.00	96	1.20	18
Kryder 5–5	2.48	80	1.16	14
Ronnse	1.94	74	1.14	14
Yamagouchi	1.78	70	1.06	6
Kryder 25–4	1.84	70	1.32	26
Rich 7–5	2.04	76	1.08	8
Rich 12–2	1.90	74	1.10	10

limitation and to find suitable methods for early detection. Some criteria used for this purpose are the *colorimetric test* (Halma and Haas, 1929), *rootstock color test* (Furr and Reese, 1946), *length/width leaf ratio* (Teich and Spiegel-Roy, 1972), *leaf shape* (Hearn, 1977), *thorn* (Spiegel-Roy and Teich, 1972), *cotyledon color* (de Lange and Vincent, 1977), etc.

Biochemical compounds have also been utilized for the same purpose, such as *infrared analysis* of oils (Pieringer and Edwards, 1965), *thin layer chromatographic* (TLC) *analysis* of *Methoxyflavonoids, Flavones and Coumerins* (Tatum et al., 1977), and *analysis of leaf and root peroxidases* (Button et al., 1976; Iglesias et al., 1974; Ueno and Nishiura, 1976; and others).

Methods for early detection of *nucellar* and *zygotic* seedlings that are comparatively reliable and relatively quick are (i) *differentiation by cotyledon color,* (ii) *polyphenol oxidase browning* and finally (iii) *isozyme pattern analysis of some key enzymes.*

1. Differentiation by Cotyledon Color

Vasquez Araujo et al. (1992) described the usefulness of cotyledon colors to differentiate the *zygotic* seedlings from *nucellar* ones. They did studies with controlled crossings. They took "Pera" sweet orange, Rangpur lime and Cleopetra mandarin as female parents and Trifoliate orange (*Poncirus trifoliata* (L.) Raf.) and its hybrids, Citrange, and Citrumelo as male parents. Seeds derived from fruits of these crosses had their embryos excised aseptically and their embryo color was noted. Fruits were collected four to five months after pollination, washed in running water, dried and the seeds removed. The seeds were then washed in a solution of detergent in water, dried, and their outer integuments stripped off. They were then treated for five minutes in 70% ethanol, and afterwards in 2% sodium hypochlorite solution for 20 minutes under a sterile laminar airflow chamber. At each stage of preparation, three successive washings were done with autoclaved distilled water. Embryos were aseptically excised under a stereomicroscope using scalpels, forceps, and dissecting needles. The inner integument was removed by a longitudinal cut at some distance from the micropylar area of the seed to avoid injury to the embryos. Separating the embryos in this way the authors found four groups of cotyledon colors—pale-white/creamy, white, yellowish-green, and light green, which indicated their *zygotic* or *nucellar* nature. On crossing Rangpur lime yielded seeds of which 1.6%, 59.1%, 33.6%, and 5.7% had pale whitish, creamy white, yellowish green, and light green cotyledons respectively when the normal color of Rangpur lime cotyledon is creamy white. Similarly crossing with Sweet orange and mandarin, they observed that Sweet orange yielded 8.5% and 91.5% pale whitish and white cotyledons respectively. Crossings with Mandarins yielded 2.0%, 19.4%, 44.9%, and 33.7% pale whitish and white cotyledons respectively, when the normal colors of the cotyledons of the Sweet orange and Mandarins are whitish and greenish respectively. Results obtained indicated that the cotyledon color of zygotic embryos was strongly influenced by the male parent, such that this character could constitute an auxillary tool in the identification of such embryos, whenever the parents in a cross possess contrasting color in their cotyledons.

While this technique may be a useful tool in research, it has little significance for nurserymen.

2. Polyphenol Oxidase Browning

When the terminal portion of growing shoots of *Citrus* is ground, the homogenates show browning. This phenomenon results from the oxidation of phenolic substrates by polyphenol oxidase enzyme. Several workers used this phenomenon to separate contrasting groups of individuals of citrus and related genera that could be divided into two phenotypic groups, *"Browning"* and *"Non-browning"* where *"Browning"* is a single gene controlled trait and dominant to *"Non-browning"*. Further, when homogenates are poured on white blotting paper, they form two concentric circular spots, the characteristics of which can be specified, enabling differentiation among individuals of the same phenotypic class. For example "Browning" species of *C. aurantium* and *C. reticulata* can be distinguished from one another by the kind and intensity of browning they produce in their inner and outer spots.

Geraci et al. (1981) tried to use this method to separate *nucellar* seedlings from *zygotic* seedlings. They took the terminal 1–3 cm of growing shoots from three-year-old seedlings and homogenized in 0.05 M potassium phosphate buffer solution (pH 7.2) at room temperature. The ratio of the buffer solution to the fresh weight of shoots was 3:1. The homogenate was then poured on white blotting paper and immediately scored for coagulation. After half an hour, they were scored for browning. When coagulation alone was scored, 10 mM potassium metabisulfite was added to the buffer solution. Pursuing this method they observed that Sour orange was "browning", and "non-coagulating" whereas Trifoliate orange was "non-browning", and "non-coagulating". Crossing these two species they obtained 91 progeny seedlings, of which 56 mono-foliate seedlings were "non-coagulating", and "browning" and identical to the mother plants. These seedlings they designated as *nucellar* seedlings. Although the method seems to be satisfactory for screening but the results did not comply with the genetic manifestation. In the segregating ratios of the "browning" and "non-browning" seedlings from the heterozygous parents, the proportion of progeny with "coagulating" phenotype was found to be very low and of the "non-browning" phenotype to be very high. The anomalies in the complementation between the two genomes modifying enzyme and substrate synthesis are obvious. In view of these anomalies, it is apparent that this method may be a sensitive and precise tool for differentiating between *nucellar* and *zygotic* progenies derived only from genetically distant parents.

Soares Filho et al. (1992) used this technique in the identification of young *zygotic* seedlings derived from polyembryonic seeds obtained through controlled crosses between parents having a distinct pattern of "browning" spots. In their experiment, the homogenate was prepared at low temperatures using liquid nitrogen and the phenotypic appearance of the spots was defined by the Sequy Color Code (Sequy, 1936), assessing their inner and outer portions, as well as their wet (visual) examination of spots (just after their formation on white blotting paper), and dry states (about four hours after the first examination). They showed that "browning" spot patterns were relatively stable among samples of the same variety, especially for the wet spots, and according to them "browning" analysis has the potential for use as a criterion of discrimination between *nucellar* and *zygotic* seedlings. But in young citrus seedlings browning analysis, in general, yielded variable results, especially in those less than two years old.

3. Isozyme Pattern Analysis

Peroxidase isoenzyme is often applied as a tool for early separation of *nucellar* and *zygotic Citrus* seedlings. Both roots and leaves were first examined for cathodic and anodic peroxidases and an inconsistent pattern in cathodic peroxidases from young leaves found. But the anodic peroxidases from young roots proved consistent.

To prepare the samples, roots were first washed in de-ionized water and dried with blotting paper. Roots (0.2–0.25 g of fresh weight) were then ground in a small pestle with 0.5 ml saline containing 20% sucrose, and extracted overnight at 4° C. The macerate was then squeezed, liquid drawn off to sample tubes, and centrifuged at 4, 000 × g for 10 minutes. Aliquots of 200 µl supernatant were then transferred to another sample tube where a drop of tracker dye (1% Bromophenol Blue in 20% sucrose) was added. An aliquot of 15 µl was layered on top of gel. Polyacrylamide gel disk and slab gel electrophoresis were conducted using 7% gel pH of 7.5. Sodium borate was used as a tank buffer (pH 8.3) for the anodic run. In addition a spacer gel (2.4% with pH of 6.9, 0.05% riboflavin, and 20% sucrose) was employed for better differentiation and greater uniformity. Gels were run at 1 mA per sample until the marker reached 85–90 mm. The tank buffer was pre-cooled to 4° C and the electrophoresis was carried out at the same temperature. Gels were stained with guaiacol and subsequently fixed in 2% acetic acid. Peroxidase bands were characterized in relation to a dark band appearing at a distance of 12–15 mm from the start. This band was analogous to the one appearing at 28–30 mm in the gel disk. Zymograms were drawn to scale with the location and intensity of each band marked. Determining the peroxidase isozymes from the root samples in this manner enabled identification of *nucellar* plants.

Geraci et al. (1981) made comparative study of different biochemical methods in order to identify genetic markers that could be used to discriminate between *nucellar* and *zygotic* seedlings. They prepared root extracts from 3-year-old Sour orange accessions using 200 mg of homogeneous 1.0 mm diameter roots and following the method described above. They prepared leaf extracts by grinding 1.5 g of leaves (approximately three months old) in a chilled mortar with 1.5 ml of 20% sucrose in the presence of acid-purified sea sand. Homogenates were then centrifuged in the cold at 6, 000 × g for 15 minutes. Polyacrylamide gel electrophoresis was performed in a vertical gel slab apparatus (Pharmacia GE-2/4). Four 8 × 8 cm slabs were run at a time under refrigerated conditions. Bromophenol Blue and pyronin G were used as markers for the anodic and cathodic runs respectively. Anodic isoperoxidases from root extracts were separated on 8% acrylamide gels, pH 7.5 as described by Manzocchi et al. (1981); cathodic isoenzymes were analyzed on 15% acrylamide gels, pH 4.1. No stacking gel was used. Zymograms for peroxidase activity were obtained by incubating gel slabs at room temperature for 20 minutes in 10 mM guaiacol, 10 mM hydrogen peroxide (H_2O_2), in 0.2 M potassium phosphate, pH 5.8; gels were subsequently fixed and stored in 7% acetic acid. They also analyzed the isoenzymes of esterase from leaf extracts on 7.5% acrylamide gels, pH 8.9. Electrophoretic analysis of peroxidases from root extracts showed differentiating banding patterns of the zymograms between the *nucellar* and *zygotic* seedlings. Esterase zymograms from leaf extracts showed better results.

Ollitrault et al. (1992) applied isozymes from bark and leaf to screen *zygotic* and *nucellar* trees. They analyzed eight enzymatic systems—Alcohol Dehydrogenase (ADH), Aspartate Amino Transferase (AAT), Endopeptidase (End), Isocitrate Dehydrogenase (IDH), Leucine Amino-Peptidase (LAP), Malate Dehydrogenase (MDH), Phospho-Gluco-Isomerase (PGI), and Phospho-Gluco-Mutase (PGM)—from leaf and bark of 25 major citrus rootstocks. Isozymes were extracted by crushing leaf (200 mg) or bark (2 cm^2) in 1 ml Tris-HCl (pH 7.2, 0.2 M) and adding Triton x-100 (0.2%) and 50 mg polyvinylpolypyrolidone. The extract was centrifuged (30,000 × g) for 20 minutes and the supernatant was immediately analyzed or conserved at (–)18° C. The enzymatic systems were studied on starch or polyacrylamide gel following the procedures described by different workers on different enzyme systems (Soost and Torres, 1981; Ollitrault et al., 1989). According to them, analyzed isozymes permit good identification of the various rootstocks and in most cases an accurate discrimination for *zygotic* and *nucellar* seedlings (or field rootstocks).

Mukhopadhyay et al. (1997) conducted isozyme analysis of peroxidase enzyme to separate *nucellar* seedlings from *zygotic* seedlings of Darjeeling orange (*Citrus reticulata* Blanco). In this 0.25 g leaf sample from seedlings more than two years old seedlings was ground with three times of its volume of the extracting buffer (0.2 M Tris-HCl, pH 6.0) in an ice-cold mortar and pestle. The homogenate was centrifuged for 15 minutes at 14,000 rpm at 4° C. The supernatant was stored at 20° C until use. Polyacrylamide gel (7.5%) was used to separate and estimate the weights of the constituent polypeptides of all the samples with the protein marker (Kaleidoscope prestained standard of the BIORAD, following the protocol of Laemmli, 1990). Electrophoresis was performed at 4° C with a BIORAD-MINIPROTEIN (II). The gel was stained with Commassie Blue solution

(0.25% Commassie Brilliant Blue, R-250, 40% methanol, and 70% acetic acid), and the excess stain was removed with a de-staining solution (40% methanol and 70% acetic acid). Zymograms were drawn to scale with the location of each band marked. As the relative mobility (Rf) of each polypeptide is proportional to its molecular weight, the corresponding molecular weight of each polypeptide was calculated by determining the Rf values (distance of protein migration/distance of the tracking dye migration) of different bands, and extrapolating the values from the standard calibration curve prepared by plotting the Rf value of the protein markers obtained on the SDS-PAGE against their molecular weights as described by Weber and Osborne (1975). The authors could differentiate between *nucellar* seedlings and *zygotic* seedlings by comparing the banding patterns of different samples. Recently Lima et al. (2000) obtained differentiation zygotic seedlings using RAPD markers.

The biochemical methods described above can separate *nucellar* seedlings from *zygotic* seedlings and are very useful in breeding programs. But they have little application in practical citriculture where the age-old method of selection by morphological characters still prevails.

Mukhopadhyay et al. (1997) examined a large number of sequential seedlings emerged from polyembryonic seeds both by appropriate nursery and biochemical studies and observed that the second seedling emerging from a seed normally appeared as a nucellar one. This method may be applied for in vitro propagation of nucellar seedlings within a very short period of time.

REFERENCES

Aubert B and Vullin G (1998). Citrus Nurseries and Planting Techniques. CIRAD, France.

Button J, Vardi A, and Spiegel-Roy P (1976). Root peroxidase enzymes as an aid in *Citrus* breeding and taxonomy. *Theor. Appl. Gen.* 47: 119–123.

de Lange JH and Vincent AP (1977). Citrus breeding; new techniques in simulation of hybrid production and identification of zygotic embryo and seedlings. Proc. Int. Soc. Citriculture, VII Int. Citrus Cong., Acireale, Italy, vol. 2, pp. 589–595.

Ebrahimi Y (1984). Identification of nucellar seedlings in seedbeds. Proc. Int. Soc. Citriculture. Int. Citrus Cong. Sao Paulo, Brazil, vol. 1, p. 14.

Furr JP and Reese PC (1946). Identification of hybrid and nucellar Citrus seedlings by a modification of rootstock color test. *Proc. Amer. Soc. Hort. Sci.* 48: 141–146.

Geraci G, Manzocchi A, Tusa N, Occorso G, Redogna L, and De Pasquale F (1981). Comparison of different methods for identifying zygotic and nucellar seedlings in *Citrus*. Proc. Int. Soc. Citriculture, Int. Citrus Cong. Tokyo, Japan, vol. 1, pp. 1–3.

Halma RF and Haas ARC (1929). Identification of certain species of Citrus by colorimetric tests. *Plant Physiol.* 4: 265–268.

Hearn CJ (1977). Recognition of zygotic seedlings in certain orange crosses by vegetative characters. Proc. Int. Soc. Citriculture, Orlando, Fla. Vol. 2, pp. 611–614.

Iglesias I, Lima H, and Simon JP (1974). Isozyme identification of zygotic and nucellar seedlings in *Citrus. J. Her.* 65: 81–84.

Lima DR (2000). Identification of citrus zygotic seedlings by using RAPD markers. *Revista rasileira de Fruticultura* 22 (2): 181–185.

Manzocchi LA, Tusa N, and Geraci G (1981). Co-relation between phenotypic and biochemical genetic markers in offspring of Sour orange *P. trifoliata. Genet. Agri.* 35: 367–376.

Mukhopadhyay S, Rai J, Sengupta RK, and Gurung A (1997). A rapid technique for the production of nucellar seedlings of Darjeeling orange. Proc. 5th ISCN Int. Cong. pp. 67–74.

Ollitrault P, Escoute J, and Noyer L (1989). Polymorphisme enzymatique des sorghos. I.Description de 11 syst'emes enzymatiques, determinisme et liaisons genetique. *L' agron. Trop.* 44 (3): 203–210.

Ollitrault P, Faure X, and Normad F (1992). Citrus rootstock characterization with bark and leaf isozymes application to screen zygotic and nucellar trees. Proc. Int. Soc. Citriculture. VII Int. Citrus Cong. Acireale, Italy, vol. I, pp. 338–341.

Pieringer AP and Edwards GJ (1965). Identification of nucellar and zygotic seedlings by infrared spectroscopy. *J. Amer. Soc. Hort. Sci.* 86: 226–234.

Pio RM, Pompeu Jr. J, and Boaventura YMS (1984). Polyembryony in 26 Trifoliate-type Rootstocks. Proc. Int. Soc. Citriculture, Int. Citrus Cong. Sao Paulo, Brazil, vol. I, pp. 24–25.

Pradhan J (1993). Polyembryony in Darjeeling orange and other Citrus species in relation to disease resistance and mass propagation of disease-free quality seeds. Ph.D. Thesis, Bidhan Chandra Krishi Viswavidyalaya, West Bengal, India.

Rajput CBS and Haribabu RS (1993). Citriculture. Kalyani Publ. New Delhi, pp. 105–119.

Randhawa SS and Bajaj BS (1958). Embryo counts and their germination in important polyembryonic citrus rootstocks. *Indian J. Hort.* 15: 61–65.

Reisfield RA, Lewis JU, and Williams DF (1962). Disc electrophoresis of basic proteins and peptides on polyacrymide gel. *Nature* 185: 281–283.

Sequy E (1936). Code Universel des couleurs. Paris. P Lechealier. (Encyclopedie Pratique du Naturaliste, 30), p. 49.

Soares Filho, W dos S, Vasquez Araujo JE, da Cunha MAP, da Cunha Sobrinho, and Passos OS (1992). Degree of poyembryony, size and survival of the zygotic embryo. Proc. Int. Soc. Citriculture, VII Int. Citrus Cong., Acireale, Italy, vol. I, pp.135–138.

Soares Filho, W dos S, da Cunha MAP, da Cunha Sobrinho, Passos OS, and Souza Jr. MT (1992). Identification of zygotic seedlings derived from polyembryonic seeds of Citrus: The use of 'Browning'. Proc. Int. Soc. Citriculture, VII Int. Citrus Cong., Acireale, Italy, vol. I, pp.139–141.

Soost RK and Torres AM (1981). Leaf isozymes as genetic markers in Citrus. Proc. Int. Soc. Citriculture, VII Int. Citrus Cong. Tokyo, Japan, vol. I, pp. 7–10.

Spiegel-Roy P and Teich AH (1972). Thorn as a possible genetic marker to distinguish zygotic from nucellar seedlings in citrus. *Euphatica* 31: 534–537.

Strasburger E (1878). Poyembryonic Jenaische Seitschr F. Naturwiss 12: 647–670.

Tatum JH, Berry RE and Hearn CJ (1977). Separation of nucellar and zygotic citrus seedlings by their flavonoids and other non-volatile compounds. Proc. Int. Soc. Citriculture, Int. Citrus Cong., Orlando, Florida, Fla, vol. II, pp. 614–616.

Teich AH and Spiegel-Roy P (1972). Differentiation between nucellar and zygotic citrus seedlings by leaf shape. *Theor. Appl. Gen.* 42: 314–315.

Ueno I and Nishiura M (1976). Application of Zymography to *Citrus* breeding. I. Identification of hybrid and nucellar seedlings in *Citrus* by peroxidase isozyme electrophoresis. *Bull. Fruit Tree Res. Stn. Japan* B 3: 1–8 (English summary).

Vasquez Araujo JE, Soares Filho W dos S, da Cunha MAP, da Cunha Sorbino AP et al. (1992). Identification of zygotic embryos in polyembryonic citrus seeds: The use of cotyledon colors. Proc. Int. Soc. Citriculture, VII Int. Citrus Cong. Acireale, Italy, vol. 1, pp. 142–144.

Weber K and Osborne N (1975). *In*: The Protein (eds. H Neurath and RL Hill). Academic Press, New York, NY, pp. 179–223.

Webber HJ (1932). Variation in Citrus seedlings and their relation to rootstock selection. *Hilgardia* 7: 1–79.

5

Citrus Production Technology

Citrus is cultivated in a very wide range of climatic conditions and land situations, from plains land to hill terraces, in over a hundred countries. It is cultivated in nearly every tropical, subtropical, and Mediterranean climatic regions of the world—from the semiarid climate of California (USA), Australia and the Mediterranean countries to subtropical to nearly tropical climate of Florida (USA), and Brazil, and, the tropical climate of Southeast Asia, extending approximately 40° N and 40° S of the Equator, concentrated primarily from 20° to 40° N and 40° S latitude.

Normally *Citrus* can tolerate wide range of soil types, from sandy loam to clay with a pH range between 5.5 and 7.5 but it is susceptible to poor drainage, salinity, alkalinity, frost, and excess boron content of the soil. The wide adaptability of different rootstocks to different climatic and soil conditions, however, makes it possible to cultivate citrus even in saline soil, frost, and other soil and climatic stresses. But *Citrus* prefers moderate temperature, rainfall, moisture, and sunlight. In general, areas with Mediterranean climate with low rainfall, cool winter, and high sunlight produce fruits with beautiful exteriors but modest internal quality. Subtropical areas with high rainfall and warm winter nights produce fruits with poor color and much exterior blemish but excellent internal quality. The climatic relationship of quality of fruit production is given in Table 5.1. Thus for application of proper production technology or, in other words, planning and programing for nursery and orchard, one has to consider not only the soil, but also the climate of the area where the citrus is to be produced and for what purpose. Planning will also differ

Table 5.1: Climatic relationship of the quality of citrus fruit products

Climatic zones	Fruit quality
(i) Humid tropics (high rainfall, warm nights)	High sugar, high juice content, thin peel, poor color, fungal blemishes
(ii) Arid deserts (low rainfall, cool nights)	Brilliant color, low sugar, high acidity, thick peel, minimum surface blemishes

with respect to the status of the planners. There may be National or State-level planning, planning for industries or private large-scale, small-scale or homeyard orchards.

NATIONAL/STATE-LEVEL PLANNING

National/State-level planning is concerned with two strategic points. Firstly, assuring the availability of fruits for the optimum per capita consumption within the country/state; secondly, producing fruits for export both as fresh and processed products. The average per capita consumption of citrus is usually 15.2 kg. The Inter-country UNDP-FAO Regional Project of the Governments of China, Indonesia, Malaysia, Philippines, and Thailand for Control of the Citrus Greening Disease drafted a plan in 1989 for citrus production in the concerned countries with the targeted consumption of 10 kg/capita/year (Table 5.2). In this plan the demand for fruits was first estimated and the deficit in production then calculated. Next the requirements of *area of nurseries, seed orchards, combined Foundation Blocks,* and *Multiplication Blocks* to be achieved in four to five years were calculated and programed throughout the concerned country. This Plan points out the basic requirements for successful citriculture, namely (i) *Seed orchards,* (ii) *Nurseries,* (iii) *Foundation Blocks* and (iv) *Multiplication Blocks.*

In any scientific citrus production system, primary items to be looked into are: (i) *Mother groves for scionic plants and seed gardens for rootstocks,* (ii) *Nuclear Block of disease-free (S-0) elite Mother or Scionic species of preference, (iii) Foundation Block of Seed plants of Rootstock species of preference,* (iv) *Foundation Block* (S-1) for budwood sources, (v) *Amplification Block of disease-free mother plant,* and *(vi) Multiplication Block* (Liner) *to produce certified budded planting materials* (S-2).

Nuclear plants (S-0) are raised under greenhouse conditions. These plants are transferred to field to raise Foundation Stock orchards or Stock Repository (S-1) as budstick sources. Modern way to raise the budstick sources is to raise Amplification Blocks (S-1). Nursery raised rootstock seedlings are transplanted to the Amplification Block and budded with the buds collected from the Nuclear or Foundation Stock under cover and with due protective umbrella. This Block facilitates the production of budsticks. Certified planting materials are produced in the Multiplication Block. This Block is raised in open field or under protective cover. Vigorous rootstock seedlings grown in the nursery are transplanted in this Block. Budsticks collected from the Amplification Block are budded to those seedlings to produce certified planting materials (S-2).

Production of budsticks by a plant is a regenerative process; so their availability is limited. The size of the *scionic nuclear/amplification block*

Table 5.2: National planning for citrus production in some South-East Asian countries (UNDP-FAO Regional Project on the Control of Citrus Greening Disease)

Parameters	Country				
	1 Malaysia	2 Thailand	3 Philippines	4 Indonesia	5 P.R. China
Population (in millions 1987–88)	16	54	61	175	1,060
Estimated demand (000 Mt)	160	540	610	1,750	10,060
Actual production, 87–88 (000 Mt)	15	640	150	600	3,700
Deficit of production (000 Mt)	(–) 145	(+) 100	(–) 460	(–) 1,000	(–) 6,350
Trees needed to cover deficit (000 units) (W)	3,100	–	9,200	20,000	1,38,000
Needed equivalent acreage of nurseries (in ha) (V)	47	–	141	307	2,123
Needed Acreage of Seed orchards (in ha) (Z)	2	–	6	13	12
Needed Acreage of Foundation Blocks (x)	5	–	15	33	230

Y: Targeted consumption 10 kg/capita/year; W: estimated production 50 kg/tree; V: Nursery density 66,000 plants/ha; Z: 1 ha seed orchards yields 150 kg seeds to be used as rootstocks; X: Combined Foundation Blocks and Multiplication Blocks to be achieved in four to five years.

will depend upon the requirement of budsticks for budding purposes. *The Seedstands of rootstocks* are only to supply seeds to raise nurseries of rootstocks, and to produce seedlings on which the budsticks are to be grafted in the Amplification and Multiplication Blocks. The Multiplication Block is to supply the *certified planting materials.* Its size will depend on *the number of seedlings required for planting or the area where planting is to be done. The seed production capacity of the rootstock species also contributes to this requirement.* Aubert and Vullin (1998) made an estimate of seed productivity in different species of rootstocks, which differs widely from species to species (Table 5.3).

Table 5.3: Seed yields of different rootstock species

Rootstock species	Seed yield/ha (kg)	Seeds/fruits	Number of seeds/kg
Citrus microphylla	440	15	5,000–7,000
Sour Orange	266	20	3,500–4,500
Trifoliate Orange	1,080	20–25	4,900–5,300
Troyer citrange	640	10–12	4,700–5,500
Citrus volkameriana	280	10–12	10,000–12,000

Once the *Foundation blocks* for rootstocks and scionic plants and *Amplification blocks* for budsticks are established and rootstock seedlings and budsticks raised, the next step in the process is obviously grafting. In the conventional system, budsticks are produced in *Foundation blocks* of scionic plants (instead of *Amplification blocks*) and budsticks are normally collected from the *scionic foundation blocks* prepared from nursery raised scionic seedlings. Nowadays, buds are collected from the disease-free microgafted plants maintained in the Nuclear Stock. These buds are grafted with the rootstock seedlings in the Amplification Block.

Whatever be the mode of seedling production subsequent steps after the production of certified seedlings include planting them in fields, managing the orchards, harvesting, and readying the fruits for the market either for direct consumption or as raw material for the processing industries. Each step or the process has its own technology that is constantly undergoing improvement through research and development.

Technologies involved in the identification and management of mother or candidate plants are the pivotal ones that in fact determine the final production and productivity. The key factors in this process are to keep them true-to-the-type and disease-free particularly free from the graft-transmissible diseases, and to produce certified planting materials.

All these processes of citrus production can be efficiently organized by individuals or private agencies having sufficient expertise and financial resources to set up the essential infrastructures and to cover operational expenditure. In developed *Citrus* growing countries, citrus cultivation is more or less well organized mostly, by nurserymen and growers; Central assistance is primarily needed for inspection and certification. In most of the potential developing countries, *Citrus* cultivation is almost unorganized and most of the nurserymen, and the growers remain at the mercy of nature. In these countries, scientific citriculture up to the production of certified seedlings needs to be planned and programed at the National level/State level, and developed infrastructures jointly by the public and private sectors, the major strategy of which would be the maintenance of the phytosanitary conditions. The Food and Agricultural

Organization (FAO) of the United Nations (UNO) and the International Plant Genetic Resources (IPGR) have framed Rules and Regulations for Phytosanitation of citrus. All the developed countries follow these Rules and Regulations to keep their nurseries and orchards healthy.

Citrus growers of the USA were the first to adopt the aforesaid citrus sanitation programs as early as 1952. Subsequently, these programs were implemented in several other countries, e.g France, Australia, Spain, South Africa, Italy, Israel, Morocco, Greece, Japan, Turkey, Portugal, and Tunisia (Aubert and Vullin, 1998). In the developing countries of South and Southeast Asia, where most of the cultivated citrus species have originated, phytosanitary rules and regulations for cultivation remain unorganized; as a result production and productivity remain far below the level of the International standard despite the fact that the area under cultivation is higher. It is thus imperative that all such countries develop strong policy for generating planting materials in accordance with the phytosanitary rules and regulations set forth for all citrus-growing regions, which further spells out the management in nurseries and orchards.

A. NUCLEAR/FOUNDATION AND AMPLIFLICATION BLOCKS

There are two approaches for the establishment of Nuclear/Foundation Blocks of Mother Groves. First of all, certified seeds of the concerned plant species can be imported following the procedures laid down by various countries for such importation. The FAO and the *International Plant Genetic Resources Institute* (IPGRI) has accredited to only a few suppliers of S-0 (completely infection-free) level Nuclear Stock (Frison and Taher, 1991). Roistacher (1991) made further additions to the list prepared by the FAO. When seeds are imported, disease-free seedlings are to be raised in the greenhouse, then grown in containers or in field specially prepared for the Foundation Block, undertaking all sanitary measures prescribed for this purpose.

Production of Nuclear stock (S-0) and Foundation Block (S-1) scionic budwoods is a highly scientific and technical matter and only institutions/ universities with the necessary infrastructures and scientific and technical manpower can do this job. For appropriate development, there should be strong linkage between these institutions/universities and the nurserymen.

1. Conventional Method for Raising Nuclear/Foundation Block of Budwoods

To establish a *nuclear stock,* the first requirement is a *candidate tree* of proper age possessing all the requisite horticultural qualities. Once the plant is identified, it is to be indexed for the graft-transmissible diseases particularly prevalent in the concerned area. Both seeds and budwoods of this plant can be collected to raise the *stocks*. Seeds are normally used only in case of remote parent trees. These seeds are to be properly treated as described elsewhere and *nucellar plants* are to be raised under insect-proof conditions. Trials are to be undertaken with these plants in isolation to test the stability of the true-to-type horticultural characteristics. Once confirmed, these *clonal source plants* are either maintained in the open field under blanket pesticide application or maintained in the insect-proof greenhouse as the sources of *clonal budwoods*. To avoid phytosanitary risks, these plants are to be regularly indexed for graft-transmissible and other diseases.

Primary needs for raising both *nuclear and foundation plants* are (i) specialized scientific team, (ii) appropriate infrastructure, (iii) appropriate greenhouse or site, and (iv) specific agronomic practices including regular sanitation and indexing.

Modern Method for Raising Nuclear Stocks

Modern method for raising *nuclear stocks* primarily involves production of disease-free planting materials by the shoot-tip grafting technique. Navarro et al. (1975) studied this shoot tip-grafting (STG) technique in detail and developed a routine procedure to obtain micrografted plants for propagation. In this procedure, they grafted shoot tips of several *Citrus* species on Troyer citrange, *Poncirus trifoliata* (L.) Raft. *Citrus sinensis* (L.) Osbeck) seedlings grown in vitro. Shoot tips considered for micrografting were composed of apical meristem and subjacent tissue with three leaf primodia, measuring 0.14–0.18 mm in height. The procedure involved six steps: *rootstock preparation, preparation of micro-bud, micrografting, care of micrografted plantlets,* and lastly *hardening* of those plants for adaptation to soil. These authors were the first to apply the STG technique in the *Spanish Citrus Improvement Programs* (Navarro and Suarej, 1977). This technique was later modified by De Lang (1978) who instead of transferring the shoot tip grafted plantlets to pots, made double grafts on the concerned rootstock seedlings. This modification made the technique more useful for raising STG plants within a very short period of time. Subsequently, this technique was adopted in the citrus production programs of various countries in particular the USA, South Africa, Israel, Cuba, Venezuela, France, Italy, Argentina, Brazil, Greece, Turkey, etc. (Navarro, 1981). In Brazil, this technique was adopted in 1981 to produce virus-free budwoods on Rangpur lime rootstock of all the major cultivars grown in that country (Santos Filho et al., 1984; Hong-Ji and Jan Yang, 1984). Starrantino et al. (1984) described the introduction of the STG technique in the sanitary improvement of citrus in Italy. Vogel et al. (1988) reported the application of the STG in Venezuela that was later utilized in the production of virus-free plants in that country (Montverde et al., 1992). In Thailand, work on the production of the STG plants was initiated in 1981 (Chartisathian et al., 1981). Chartisanthian and Tontyporn (1990) standardized the technique to produce the STG plants of *Citrus reticulata* Blanco on two rootstocks, Cleopetra mandarin (*Citrus reshi* Hort. ex Tanaka), and Somsa (*Citrus aurantium* L.).

This technique has likewise been standardized in several other South Asian countries, in particular Taiwan and the Philippines but still remains confined to the concerned laboratories and has yet to be integrated into their citrus production system (Gonzales et al., 1991). In China, on the other hand, the STG technique has been introduced for the establishment of the propagation system of disease-free citrus trees. In India, Mukhopadhyay et al. (1997) standardized this technique to produce tristeza- and greening-free Darjeeling orange (*Citrus reticulata* Blanco) on Rough lemon and Rangpur lime rootstocks.

1. Preparation of Rootstock

The preparation of rootstock involves selection of the species, collection of seeds preferably from a clonal plant, seed treatment, decoating of seeds, sterilization of them, germination in vitro under complete darkness, and growth for a specific period depending upon the species. Freshly collected seeds are first treated with a hot water dip for 10 minutes at 52° C followed by a short dip in cool water.

Treated seeds are then decoated and sterilized with 0.5% sodium hypochlorite solution containing 2–5 drops of 0.1% Tween-20 for 10 minutes. They are then thoroughly rinsed in sterile distilled water. Sterile decoated seeds are germinated in MS (Murashige and Skoog, 1962) medium, and allowed to grow at 25–30° C for 14–21 days depending on the species.

2. Preparation of Microbud

Budsticks (5–10 cm long) are collected from disease-free *candidate* or *clonal* tree (not more than 10–15 years old). Large leaves are stripped off. The sticks are then surface sterilized with 0.25% sodium hypochlorite solution containing 0.1% Tween-20 for 15 minutes and thoroughly rinsed with sterile distilled water. The sterile budsticks are then placed in large sterile glass tubes containing moist sterile sand at the bottom or in sterile agar stabs at 32° C with 16 h daylight of 1200-lux intensity for bud breaking. Freshly emerged buds (normally appear within 10–14 days) are aseptically removed and the shoot tips (apical meristem and adjacent tissues with leaf primordia, < 0.2 mm long) or microbuds are dissected in sterile conditions using a binocular stereomicroscope placed into a laminar airflow cabinet.

3. Grafting of Microbuds

In vitro grown rootstock seedlings are aseptically taken out from the culture tubes in laminar airflow cabinet and decapitated, leaving about 1.5×10^{-2} m of the hypocotyl. An incision is made (Vertical, T) on the hypocotyl through the cortex to the cambium with a sterile razor blade and excised shoot tip carefully placed inside the incision. Each micrograft is then aseptically cultured in liquid MS medium containing thiamine-HCl (0.1 mg l^{-1}), nicotinic acid (0.5 mg l^{-1}), and sucrose (45 gm l^{-1}). Sterile folded supportive filter paper is placed on the nutrient solution; the center of the platform is perforated before placing it inside the culture tube for inserting the root of the rootstock seedling.

4. Growth of Micrograft

Culture tubes containing the micrografts are incubated at 27–32° C giving specific light conditions depending on the rootstock-scion combination. Normally they are kept in complete darkness followed by low light (750 W) 16 h day^{-1} for 15 days and high light (1500–2000 W) for 16 h day^{-1} for 39 days. The extent of success of micrografting varies from species to species of the rootstock. It also depends upon the skill of the technician doing the grafting.

5. Hardening of Micrograft and Raising Nuclear Stock

The micrografts from the culture tubes are transferred to suitable containers (black polyethylene bags) containing Soilrite and placed in an acclimatization chamber for 7–15 days. They are then transferred to plastic pots/bags containing sterilized soil mix and kept in the greenhouse for six months. The grafts are then indexed against prevalent graft-transmissible diseases. The plants should then be ready for container-growing or transplanting in orchards. These micrografts would require a long period for bearing. Double grafting of the micrografts onto the concerned rootstock significantly reduces this time period. The micrografts from the tubes can be directly regrafted on the greenhouse-grown 1-2-year-old rootstock seedlings. The age of the micrografts for double grafting also depends on the rootstock species. These double grafts are covered by polyethylene bags for seven days in the greenhouse and allowed to grow for six months, then indexed against the diseases. These plants will then be ready for planting as *Nuclear Budwood (S-0) Repository* inside a glasshouse.

Advantages of Shoot Tip Grafting Technology

Application of this technology frees the budwood sources of occasional viroids, viruses or prokaryotic organisms that may have been present priorly in the buds. As the distal part of the meristemic shoot is devoid of vascular tissues, it remains potentially "clean". This technology protects the genetic integrity of the budwood source. Shoot-tip grafted plants show no juvenility and the first bearing may occur even earlier than the conventionally grafted plants.

It is noteworthy, however, that the shoot-tip grafting technique does not provide 100% assurance for disease-free plants. There may be pathogens other than graft-transmissible ones. To avoid these chance pathogens, thermotherapy is also recommended along with the shoot-tip grafting. Moreover, this technique is almost an art and its performance, including the extent of the elimination of diseases depends on the efficiency of the operator. Once produced the *nuclear plants* are to be regularly indexed to retain confidence in their health status.

Budstick Amplification Block

Considering the limitation of the conventional Foundation stock to provide sufficient budsticks, *amplification blocks* are raised in protected field to produce sufficient number of budsticks. To raise these blocks, disease-free buds (S-0) collected from *nuclear stock,* are grafted on vigorously growing rootstock seedlings to obtain sufficient budsticks. It is recommended that budded plants be grown in high density with a spacing of 3.5–4.0 m accommodating approximately 570 plants ha^{-1}. Rate of fertilization should be higher at least 1.5 times more than that applied in fruiting orchards. Nitrogen is normally applied earlier in the season to accelerate the physiological growth of the bud-bearing twigs. Blanket pesticide applications are made more frequently to avoid epidemiological risks. As the buds are collected before fruiting, residues pose no problem.

In these blocks rootstocks may also be planted at a density of 8000 plants ha^{-1} and pesticides are regularly sprayed at an interval of 10 days to protect them from pests and diseases.

These plants are allowed to produce budsticks up to three years. The planting materials raised within this period are sold out or transplanted and left-overs destroyed to raise fresh planting materials. Budwoods can be collected three times a year, each time over a period of three months separated by a resting period of 30 days. Depending on the species, each plant produces at least 45 buds in the first year and 90 and 135 buds in the second and third years respectively (Aubert and Vullin, 1998).

Harvesting of Budsticks

Budsticks after collection are defoliated by cutting the leaves at the base of the petiole. From 20–25 budsticks are then bundled, wrapped in moist cloth, and tagged indicating the number of the source tree and the name of the scionic varieties/cultivars. The bundles are then soaked in a fungicide solution (0.5% Benomyl mixed with 0.5% Captan) and dried under shade.

Transportation and Conservation of Budsticks

Budsticks thus prepared can be transported a long distance after waxing (dipping in liquid paraffin) and putting them in plastic bags with appropriate labels inside and outside the bags. These bags can be carried to a distance that can be covered normally within 24 h. Bags in cold storage (10° C and 75–80% relative humidity) can be conserved for two to three months. Budsticks can also be conserved for more than five months but less than eight, by storing them at 4.5° C.

Candidate Seed Stand and Raising Seedlings of Rootstocks

Rootstocks are more or less of equal importance if not more as scionic plants in citriculture. They provide nutrition and keep the scionic plants standing and productive under all biotic and abiotic soil stresses. They also influence several horticultural characters of scionic plants related to their production and productivity. The characters they normally influence are juice quality, fruit maturity, on-tree storability, postharvest shelf life, tolerance to frost, drought, flooding, pests, and diseases, and adaptability to different soil texture, pH, calcium carbonate content, etc. Rootstock in fact forms the foundation for the performance of scionic plants. Thus the basic requirement for establishing any citrus orchard is selection of proper rootstock with high polyembryony, selection and maintenance of the candidate tree which is true-to-type (nucellar seedling), and freedom from all pests and diseases.

Seedmother trees are normally obtained from seeds or by grafting. Seed plants show juvenile characters, in particular thorniness, and late fruit bearing capacity. By grafting, these characters can be avoided. Grafting can be done on the same variety or varieties well adapted to the soil and climate where the orchards are to be raised. Raising seed orchards from seeds has been dealt with separately.

Raising of Seedlings: Indoors

Seedlings are raised from stored or fresh seeds. All seeds should be disease-free. Before planting, seeds are to be suitably treated to avoid seed-borne pathogens those are carried both internally and/or externally. Generally fungi such as *Colletotrichum, Diplodia, Alternaria, Penicillium, Aspergillus, Geotrichum, Rhizoctonia,* etc., and such bacteria as *Xanthomonas, Pseudomonas,* etc., are externally seed-borne, while several viruses and viroids are internally seed-borne. To remove the pathogenic contaminants from seeds, they are to be surface sterilized with 0.5% sodium hypochlorite solution containing 0.1% wetting agent like Tween-20 for 10 minutes. In addition, seeds may be treated by hot water at 52° C for 10 minutes to eliminate internally borne pathogens. The treated seeds are then surface-dried under shade and dusted with Thiram (75%) or dipped in 1.0% solution of B-hydroxy quinoline sulfate and air-dried. These seeds are then packed in small polyethylene bags, pieces of moist tissue paper placed between these bags, and the small bags are then placed inside larger ones that are stored at 5°–6° C.

However, freshly collected seeds are always preferred for propagation. In epidemiological risk areas seedlings should be raised in an insect-proof greenhouse. The seedlings may be nursed inside the greenhouse or transferred to suitably prepared nursery beds. To achieve a healthy orchard, plants must be cared for from childhood or, in other words, from the nursery.

Inside a greenhouse, seedlings are normally raised either in Root Trainer Cells (10 cm unit) or in wood/plastic boxes. In a Trainer, seeds are planted in cells with at a minimum depth of 10 cm. Each cell holds the root system of each seedling as a separate entity and provides the opportunity for potting up each seedling separately without root damage; it also prevents competition between seedlings for available nutrients and imposes uniform light for each seedling. Raising seedlings in Trainers also provides better opportunities to discard zygotic seedlings, and makes uniform grading of seedlings possible. It further facilitates watering control by allowing inspection of seedling roots from time to time and checking on the penetration of water. Once prepared, the carriers of single cells are kept on benches in the greenhouse. The Trainer Cells or the containers made of wood/plastics are always to be sterilized before use. Wood containers are normally dipped in copper naphthenate solution. The plastic containers may be sterilized by free steaming; drainage holes are to be made in these containers before sterilization. The size of the containers normally used in propagation is 40 × 40 cm or 40 × 20 cm and 14 cm deep.

The sterilized containers are to be filled with very carefully prepared soilmix.

Preparation of the Soilmix

The soilmix normally contains 50% peat moss or its substitutes (Annexure 3) and 50% sand with macro- and micronutrients. Bark from both hardwood and coniferous species may be substitutes for peat moss. Pine bark/sawdust, etc. may improve the nutrient level of the mixture. The size of the sand particles in the "mix" should be of 0.05–0.5 mm. Care should be taken to avoid contamination of sand with clay or limestone. Sand may be substituted by vermiculite or perlite in a proportion of 1/2 peat, 1/4 vermiculete, and 1/2 or 1/3 of each other ingredient. The components of the "mix" are to be thoroughly mixed in an electrically operated concrete mixer.

The weighed quantity of macronutrients (superphosphate, calcium carbonate, and magnesium carbonate) are sprinkled on the peat moss positioned on the apron of the mixer, then added to the fine sands in the mixer, and the contents thoroughly tumbled. The micronutrients [copper sulfate ($CuSO_4$, $5H_2O$), zinc sulfate ($ZnSO_4$, 36% Zn), manganese sulfate ($MnSO_4$, 20% Mn), ferrous sulfate ($FeSO_4$, $7H_2O$), molybdenum (Mo_7O_2, $4H_2O$), and boric acid (H_3BO_3)] are weighed and mixed together, dissolved in water and then poured into the turning mixture. After 20 minutes of tumbling, the "soilmix" is removed to a soil-sterilizing unit for steaming, the time of which depends upon the quantity of steam produced, which again depends on the size and capacity of the boiler. Normally free steaming is done for 15 minutes at 83° C or 1–2 minutes at 100° C. The sterilized "soilmix" is then used to fill the containers for raising seedlings.

Filling Containers with the "Soilmix"

The sterilized "soilmix" is placed in containers and compacted by means of a metal tamper, keeping the level of the soil to 3 cm from the top. A planting board made of thin masonite or plastic with 1.5 cm diameter holes drilled 2.5 cm apart is used to uniformly plant the seeds. After seeding, the planting board is removed and the seeds covered with soil (1.0 cm) and lightly tamped. Watering is done with suitably treated water obtained from water treatment plant. Water can be applied using the soft spray sprinkler nozzle on a hose or a watering can with a perforated sprinkler head. Seed containers or seedling trays kept at room temperature during summer or in warm conditions during winter until the plants reach transplanting size (8–15 cm). This period depends on the variety and climatic conditions, and varies from 25–51 weeks. The seedlings are to be critically examined once a week for off-types, gametic, and nonnucellar variants. Separation of nucellar seedlings is a key process in primary citriculture. Working or field separation is normally done by visual observation. Nucellar seedlings should be of medium and uniform height, more branched and less thorny.

Raising Seedlings in Outdoors in Seedbeds and Liners

To realize healthy seedlings in outdoor conditions, utmost care must be taken in the preparation of seedbeds and their management. Operators are to take proper precautions by disinfecting their shoes, clothing, applicators, etc. before undertaking any operation in the seedbeds.

Beds should not be located near trees, hedges, shrubs that may inhibit air movement across the beds. Beds should also be far removed from infected plants (if any), and hosts of vectors that may carry viruses and greening bacteria, and other fastidious pathogens. Aphids and psyllids mostly carry them. The list of common hosts of psyllids is annexed (Annexure 4). To select a site, soils virgin to citrus should be of first priority. If old citrus soil is used, it is to be fumigated before planting. Rotational cropping combined with fallowing prevents several soil problems. The preferred rotation is a three-year cycle of legumes/cereals as noncitrus crops.

Seedbeds should be elevated above the surrounding soil and path areas. Materials should be added while preparing the beds to ensure that the soil is very free draining. The mixture should be made uniform throughout and contain a high percentage of sand. Provision should be made to screen beds from direct overhead sunlight and to protect them from heavy rains. There should be provision for at least 50% shade. Seed rows in beds should be at least 10 cm apart, preferably even wider. The space between each seed in the row should be 5 cm or more. Seeds should be planted no deeper than 2 cm. It is preferable to make grooves, plant the seeds in them, and cover the bed surface with coarse sand (1.0 cm). The ratio for the surface area and seed and transplant bed is 1:20 or 1:25. A surface area of 500 m^2 can supply seedlings for 1 ha planting area (30,000 plants). Keeping the rotation and fallow land requirements in mind it is recommended that the total surface area be six times more (for 1 ha planting area, the total nursery area should be 6 ha).

Seedbeds are disinfected at least 4–5 months prior to seed planting. This can be done by thoroughly wetting the soil with vapam (sodium N-Methyl-di-thio-carbamate dehydrated, 1.2 l active ingredient per 5 liter of water). But this treatment is effective only at low temperature (12–15 C). If the soil is treated, it is to be rolled for compaction, watered, and then tested for the presence of any toxic residue (watercress test).

"Solarization" is the alternative low-cost, environmental-friendly method for soil disinfection. After preparing the seedbed, the soil surface is punctured to make holes (1.5 cm wide and 5.0 cm long) using a board for making such holes. A transparent plastic sheet (70 µm thick ethyl acetate vinyl) is placed over the suface of the seedbed. The punctures help diffuse heat into the soil. Planting can be done 2.5–3 months after solar exposure. Recently Ghini et al. (2000) developed a solar collector to promote complete eradication of *Phytophthora* from the soil substrate. It has been claimed that 20 solar collectors can eradicate the fungus from 100 m^3 soil substrate. The estimated annual cost/m^3 soil would be $ 0.42 to $ 0.32.

Before planting the soil may be artificially mycorrhized (compost, forest litter, river sand, etc. collected from the noncitrus area). Formulated mycorrizal cultures may also be used where available. Basic dressing of the soil ha^{-1} is done with 80 tons well-composted manure, 0.4 tons triple superphosphate (45% P_2O_5), 1.0 ton dolomite (in soil with pH 5–6), 0.5 ton potassium sulfate (50% K_2O), and captafol or quintozene at the recommended dose. To sow seeds the soil surface is depressed (2.0 cm deep) with a stick every 20 cm. The rate of seed sowing is 150 seeds/m^2. The beds are then covered with a 2.0-cm layer of river sand or peat moss.

Immediately after planting the seeds, a thorough fungicidal drench should be applied to avoid soil-borne pathogens, if the soil was not properly treated before seed planting. Seedbeds are to be properly irrigated, preferably by a sprinkler, at the rate of 45 m^3 ha^{-1} every 3–4 days for half an hour in the afternoon until the seedlings emerge. As the seedlings become 10–15 cm tall, irrigation should be extended (120–130 m^3 ha^{-1} for 1.5 hours, once in a week). In between two water sprays the topsoil is lightly hoed. Bordeaux mixture is applied before the rains to avoid bacterial infection. In case of scab infection and powdery mildew, protective sprays of methyl thiophanate (to prevent scab) and triadimefon or dinocap (for mildews) are to be applied at the recommended dose.

Attention to seedling uniformity is essential. The less bench-root system should be allowed to ensure uniform competition for light in seedbed conformities and proper grading done at planting-out time to obtain true-to-type plants. Seedlings smaller than 15 cm (except for Trifoliate orange) are to be discarded. Once the seedlings reach the proper size, they are lifted, detopped (above 25–30 cm), their taproots are cut off at 10–15 cm, and defoliated. They are then treated with an anti-transpirant. Roots are placed in a semiliquid mixture containing 60% soil virgin to citrus and 40% rich composted manure mixed with a rooting hormone. These seedlings are then lined out in the *sandwiched foundation block* or *multiplication block* for budding. In an intensive planting system, the quantum of planting may be done with a spacing of 0.75 m between rows and 0.33 m between plants (40,000 liners per ha). Irrigation and fertilizer requirements will also be high. Liners in high density planting should receive irrigation (sprinkler) every 10 days (400–500 m^3 water ha^{-1}). Other operations include hoeing, weeding, and fertilizer application, once in summer (before rains) and another in autumn (post-rainy period) at the rate of 40 kg ha^{-1} N and 20 kg ha^{-1} N respectively in the form of ammonium nitrate in acid soil, and ammonium sulfate in neutral or alkaline soil. The liners are to be desuckered, and the lateral branches removed so that the rootstocks develop with a straight stem.

Raising Seedlings in vitro

Raising seedlings in Trainer Cells, containers, and seedbeds, is rather time consuming and the extent of germination remains limited, whereas in vitro raising of seedlings is very rapid and ensures almost 90–100% germination. To grow seedlings in vitro, seeds are decoated and sterilized by treating with 0.5% sodium hypochlorite solution adding 2–3 drops of 0.1% Tween-20 for 15 minutes, then thoroughly rinsed in sterile distilled water. These seeds are then put on an MS medium for germination, and growth, keeping them at 27–30 C. Seedlings of polyembryonic seeds start to appear within 5 days and continued to emerge up to 21 days depending upon the status of polyembryony. When four seedlings emerge from a seed, the 4th seedling normally does not survive and growth of the 3rd seedling is very slow.

The growth of the 1st and 2nd seedling, on the other hand, is comparatively higher (Table 5.4). As stated before, the 2nd seedling normally becomes a nucellar one.

Table 5.4: Growth of sequential seedlings from polyembryonic seeds of Darjeeling orange (*Citrus reticulata* Blanco) in vitro (Mukhopadhyay et al., 1997)

Sequence of germination	Days taken for root appearance	Days taken for shoot appearance	Growth of seedling after 30 days of emergence of shoot		
			Root length (mm)	Shoot length (mm)	No. of leaves
First	4–10	8–15	46–98	11–45	2–4
Second	4–18	9–23	11–75	10–30	2–4
Third	10–21	17–23	2–15	5–30	2–4
Fourth	20–21	22–25		did not survived	

These seedlings are separated out after 21 days, put to Trainer Cells inside a greenhouse for further screening for uniformity. They are then transferred to small polypacks filled with sterilized "soilmix" and allowed to grow for 5 months. The nucellar seedlings are then transferred into bigger polypacks or suitable containers and allowed to grow inside a greenhouse for 6 months or until they reach the grafting stage. Alternatively, they may be transplanted to liners.

This procedure produces very healthy, apparently disease-free and nematode-free seedlings but it needs to be standardized in different geoclimatic conditions (Mukhopadhyay et al., 1997).

Raising Seedlings by Tissue Culture

Mukhopadhyay and Chaturvedi (2000) standardized the techniques for producing clonal seedlings of both Rangpur lime and Rough lemon by shoot tip proliferation and nucellar embryogenesis. It was possible to regenerate the plants ready for hardening within 80–90 days and the regenerated plants could be transferred to containers or field nurseries within 130–150 days.

They found tissue-culture plants superior to those raised from seeds, with respect to vigor, morphometric characters and root system. The budding efficiency of tissue-culture rootstocks and seed-raised rootstocks was found to be more or less the same. Commercial firms are now supplying tissue-culture rootstocks, for example, Tyford International in Florida (USA) and Delbard Ltd. (France) (Aubert and Vullin, 1998).

Raising Seed Gardens

Seed gardens are required to obtain the requisite quantity of seeds for raising rootstock plants for propagation for such gardens, seedlings are initially raised according to the methods described above and then transplanted to the field. While undertaking this process due attention must be exercised against citrus canker. All the conventional rootstocks are susceptible to this disease. Attention is also necessary to avoid soil-borne pathogens, in particular *Phytophthora* and nematodes. Furthermore, maintenance of true-to-type seedlings needs to be ensured.

Normally plots are raised in isolation with adequate windbreakers. A planting density of 6 m × 4 m is followed, resulting in 400 trees ha^{-1}. Blocks are to be specially designed to check cross-pollination and to maintain purity of the type. After transplantation, the usual fertilization, irrigation, blanket pesticides application, and basal pruning are pursued. As the plants start blooming, it is recommended to place beehives in each block to increase the chances of intra-block pollination. In a seed garden, plants start to produce fruits after 5–7 years but full bearing is normally not reached before 15 years. In grafted seed plants, however, uniform fruiting can be obtained within 6–7 years. Each tree normally produces 50 kg of fruits excluding Trifoliate orange that produces less fruits. Each kg fruits normally yield 3,000–6,000 seeds depending upon the species, excluding *Citrus volkameriana* that produces more seeds per kg fruits. Only fully matured fruits are plucked. Seeds collected from immature fruits normally do not germinate. Plucking time is decided when the mature fruits start dropping, but dropped fruits are not to be considered for seed extraction as they may be contaminated by soil-borne pathogens.

Extraction of Seeds and Postextraction Conditioning

Extraction of seeds from fruits is done either manually or by machine. After extraction they are treated with hot water at 52°C for 10 minutes to disinfect them followed by a short dip in cool sterile water. They are then dipped in 1% solution of hydroxy quinoline for three minutes. Alternatively, after hot and coldwater treatment, they are dried in a shaded area that is well ventilated and are regularly tumbled. Dried seeds are dusted with 75% Thiram powder and stored.

In manual extraction the seeds are transferred into polyethylene bags, weighed, properly labeled and stored at 3–4°C and 60–90% relative humidity. In case of machine-extracted seeds, they are normally stored in vacuum-packed bags, kept at 5 °C temperature and 85% relative humidity after proper labeling.

Grafting Budsticks on Rootstocks

Once the budstick and rootstock Multiplication/Amplification Blocks are established, the next obvious step is the grafting of budsticks on rootstock seedlings. In fact, grafting is the beginning of the actual production system and for successful grafting the following particulars need to be properly addressed: (i) *compatibility between budstick and rootstock;* (ii) *source of budsticks, its age, position on the twig at the time of collection;* (iii) *health of the budstick source plant at the time of its collection;* (iv) *postcollection sanitary treatment;* (v) *condition of the rootstock, its age, size, whether container-grown or field-grown, its nutritive status, type and fertility status of soil, if field-grown, health status,* etc.; (vi) *season and climate at the time of budding operation;* and (vii) *sanitation* (*grafting/budding equipment and utensils*).

Rootstocks and scions should have perfect compatibility. The compatibility status differs with different combinations (Table 5.5).

Table 5.5: Compatibility of different scionic plants with different rootstocks

Rootstock	Compatibility status
Sour orange	All varieties except some Satsuma And Kumquats
Trifoliate orange	All varieties
Troyer citrange	All varieties
Carrizo citrange	All varieties
Cleopetra mandarin	All varieties
Citrus volkameriana	Limes and Lemons
Citrus macrophylla	Limes and Lemons
Trifoliate orange-Flying Dragon	Grapefruits, Tangelos, Limes
Citrumelos	Grapefruits, oranges, mandarins

Freshly collected budsticks from a young twig of the apical region of a healthy source plant always do better than stored budsticks. Sanitation of the budsticks is always to be looked after before using them. Rootstocks should be healthy, straight and without any branch. The age of the rootstocks may vary from nine months to two years depending upon the location, particularly altitude, as the growth of plants at high altitude is always slower than that in the plains; similarly the growth rate is faster in tropical, and subtropical conditions than in temperate climate. Growth is always better under temperature, relative humidity and shade-controlled greenhouses. In any case, the stage when the seedling reaches pencil diameter (8 mm) is the suitable location-specific age of the rootstock for grafting/budding. Success of grafting/budding, moreover depends on the season in open fields and shaded nurseries. Between grafting and budding, the latter is preferred in citrus propagation. Success of budding is generally related to the growth pattern of the rootstock and flushing time of the scionic plant. The rootstock should be in an active growth stage and the cambium cells actively dividing so that the bark separates readily from the wood. It is also necessary that well developed buds of the desired cultivars be available at the same time.

Normally these conditions occur at three different times in a year. In the Northern Hemisphere, these periods are late July to early September (Fall budding), March and April (Spring budding) and late May and early June (June budding). In the Southern Hemisphere, similar periods would be late January to early March (Fall budding), September and October (Spring budding) and late November and early December (so-called June budding). It is, however, the Fall budding which is most productive.

Once all these physical, biological and climatic parameters are properly taken care of, next comes the actual operation. In spite of all these duly attended factors, the success of which depends upon the skill of the operator. This may be as high as 100% at the hand of an efficient technician or nurseryman and as low as 30% for an inefficient hand.

There are various methods of budding, among which T-budding (shield budding) is preferred in citrus propagation. The T-bud designation arises from the T-like appearance of the cut in the stock whereas the name "shield bud" is derived from the shield-like appearance of the bud piece at the time of its insertion in the stock.

For budding, it is necessary to start the prebudding preparation of the rootstock at least three months before the actual operation. Seedlings are to be desuckered and the lateral shoots and thorns removed up to 35 cm above ground level to keep the surface smooth. Fertilizers and irrigation are also to be provided to make the seedlings vigorous. If necessary, pesticides are also to be applied to keep the seedlings free from pests and diseases.

In starting the operation, all sanitary precautions are taken. Then incisions are made on the smooth surface of the rootstock. The first cut may be longitudinal or vertical. Normally, a vertical cut (2.5 cm long) is made first, then the horizontal one is made through the bark (approximately one-third distance around the stock); the knife is given a slight twist to open the two flaps of the bark. After making the T-cut, the bud is prepared for insertion. Starting about 1.2 cm below the bud, a slicing cut is made under and about 2.5 cm beyond the bud; about 2 cm above the bud, a horizontal cut is made through the bark and into the wood, permitting the removal of the bud piece (2.0 cm long and 1.0 cm wide); if necessary, a spatula may be inserted underneath the bud to facilitate its detachment. The shield piece thus removed is then inserted into the T-cut by pushing it downward under the flaps of the bark of the stock until the horizontal cuts on the shield and the stock are even. The bud union is then properly tied with some wrapping material such as rubber budding strips, parafilm tape or plastic ties of polyvinyl chloride film (10 mm wide). The stock is then decapitated slightly above the budded portion and kept under polyethylene cover for a week. The strip/tape on the union is removed after the union has healed.

In citrus budding, often *inverted T-budding* is practiced, especially in rainy localities. This method is basically the same as the conventional T-budding. In inverted T-budding, the incision in the stock has the transverse cut at the bottom rather than at the top of the vertical cut and the shield piece is removed; while removing the shield piece from the budstick, the knife starts above the bud and cuts downward below it. The shield is removed by making the transverse cut 13–19 mm below the bud. The shield piece containing the bud is inserted into the lower part of the incision and pushed upward until the transverse cut of the shield meets that made in the stock. Budded trees grown in nurseries become ready for transplanting in orchard locations within 2 to 3 years from the time of sowing seeds of rootstocks in the nurseries.

In case of budding of container-grown rootstocks, the plants reach the transplanting stage within 14 to 18 months from the time of seed sowing. Hartmann et al. (1993) have described in detail different methods of grafting/budding used in propagating different types of plants.

Orchard Planning

Certified planting materials (desirable scion budded on desirable rootstock) raised in nurseries (Multiplication Block/Liners) are subsequently transplanted in orchards that are also to be properly planned. The first and foremost item in this regard is the " site". If the site is not suitable or properly organized, the orchard may fail to produce the desired output. The site should always be located far away from any existing infected orchard. Land-use planning is to be done according to the physical, chemical and biological nature of the land, i.e, hilly slope, flat or marshy land. After fixing the site, all the physical, chemical and biological aspects should be reevaluated.

Orchards do not grow well in poorly drained, highly acidic or calcareous soil and white sands devoid of organic matter. Citrus plants prefer sandy to clay loam soil with a pH 5.5 to 7.5. Before planting, the nutrient status of the soil, in particular the chloride, iron, boron, and calcium content should be determined. Newly cleared soil may be deficient in phosphorus and other essential nutrients. Old grove land may contain a high level of copper, pathogens and nematodes. In low rainfall areas, provisions may be made for adequate irrigation.

After selecting the proper site, the land must be made suitable for planting. The preparatory processes include the clearing and burning of existing vegetation; if the water table is near the surface, an extensive drainage system is installed; in hilly areas the site should be located on the south face of the hill; plantations are to be in contoured lines and the terraces properly plowed and shaped with topsoil before planting; in lowland areas, planting should be done on large individual earthing up butts to avoid waterlogging; earthing up is continued for connecting citrus trees on plantation lines. There should also be suitable windbreakers against trade winds to protect the plantation. Once site selection and preparation are completed, the orchard development work can be taken up.

Basic chemical dressing is done in acid soils (pH < 5.5) with slags (2.0 tons ha^{-1}), and limestone (1.0 tons ha^{-1}) mixed with dolomite if necessary. In neutral or alkaline soil, application of limestone is not required and slags are substituted with superphosphate, and ammonium sulfate. Once the site is ready for planting, proper layout is done. Contour lines are drawn with a gentle slope of 0.77% to allow evacuation of excess rainfall. After completion of the layout, individual holes 80 cm × 80 cm × 80 cm are dug to plant each tree. The budunions are kept well above the ground level. The trunks of the young trees are painted with a mixture of white vinyl paint plus 90% copper sulfate. Protective netting is also placed around each plant to prevent rodents and other animals from damaging the plants.

After planting, orchard management is done according to a specific schedule for watering, fertilizing, green manuring, intercropping, weeding, pruning, etc. During the few months after planting, only care for irrigation is necessary. When the young plants start producing shoots, nitrogen dressing at the rate of 25–30 g is applied followed by watering. In the case of low soil pH, phosphorus and potassium fertilizers are applied in the second year. In acid soil, limestone is applied every 2 years. Nitrogen fertilizer is applied around each tree during the first 3 to 4 years, 3 times a year, 1/4 in the early spring, 1/4 two months later, and 1/4 during fruit drop. Between the 4th and 7th years, nitrogen is applied twice a year to rows at the drip lines of the canopy, 1/2 just before blooming, and 1/2 during fruit drop. The actual program for fertilization is to be drawn up, however, by periodic analysis of leaf and soil.

Additional fertilization is done through green manuring (40 kg N ha^{-1}). These plants are sown in between rows in the orchard 3–4 years after planting. Common green manure crops are lupine (1st year, 120 kg seed ha^{-1}); mustard (2nd year, 20 kg seed ha^{-1}) or Chinese radish (18 kg seed ha^{-1}); field bean (3rd year, 120 kg ha^{-1}); and mustard or Chinese radish (4th year).

To achieve good productivity, citrus plants are pruned annually for better light distribution to the entire canopy. The terminal shoots of young nursery trees are trimmed 60 cm above the ground. In the beginning of the second year, 3 to 5 well grown scaffold branches are kept and the rest lateral branches are trimmed. During the next growing cycle, 60 to 80 cm long shoots produced by the scaffold branches are trimmed to have 6 to 8 branches, well distributed in tiers, giving a spherical canopy. Pruning is also continued in adult trees, where dead wood, bushy vertical shoots, weak central fruiting twigs, and water shoots are thinned. The cuts after pruning are smeared with a protective healing mastic. Pruning utensils are disinfected after each operation.

While managing a citrus orchard, intercropping may also be done for additional income with compatible crops such as vegetables, beans, peanut sown in between tree rows.

Integrated Pest and Disease Management

Management of pests and diseases starts from the stage of site selection for the nursery and orchard. These are to be located 3.0 km away from any infection/infestation source. Both the nursery and orchard are to be protected by windbreakers as most of the common pests, vectors, and pathogens are normally brought by wind from distant sources. The windbreakers are planted perpendicular to the trade winds at least 8 m away from the first line of *Citrus*. *Leucaena glauca* and *Erythrina fusca* are normally grown as windbreakers in humid tropical areas.

The next step in management is selection of rootstocks having acceptable tolerance to the diseases and pests prevalent in the area. Plantation is initiated with disease-free planting materials produced by shoot tip-grafting or certified planting materials. Before planting, statistically representative samples are tested for the presence of any pathogen or pest, including nematodes. Special attention is given to vascular-borne pathogens. In spite of undertaking all preventive measures, chance infection/infestation of certain pathogens/pests may occur. Methods to be adapted to control such diseases and pests are discussed in respective chapters later.

REFERENCES

Aubert B (1989). Preventing citrus debilitating diseases for profitable crops. *In*: South East Asia Inter-country UNDP-FAO Regional Project of China, Indonesia, Malaysia, Philippines, and Thailand for the control of citrus greening disease UNDP-FAO RAS/86/022.

Aubert B and Vullin G (1998). Citrus Nurseries and Planting Techniques. CIRAD, France.

Chartisanthian J and Tontyporn S (1990).In vitro culture for the production of disease-free citrus in Thailand (1990). *In*: Rehabilitation of Citrus Industry in the Asia Pacific Region. *Proc. 4th Asia Pacific Int. Conf. Citriculture* (eds. B Aubert, S Tontyporn, and D Buangsuwon) Chiang Mai, Thailand UNDP-FAO Regional Project RAS/86/002, pp. 69–75.

Chartisanthian J, Tontyporn S, and Sopitkun S (1981). Shoot-tip grafting of Ma-Nao (*Citrus aurantifolia* Swingle). *Ann. Res. Report. Dept. Agric.*Thailand.

Chung Ke (1991). The present status of citrus huanglungbin and its control in China. *In*: Rehabilitation of Citrus Industry in the Asia Pacific Region. *Proc. 6th Int. Asia Pacific Workshop on "Integrated Citrus Health Management"* (eds. Ke Chung, Shamsudin and B Osman), Kualalumpur, Malaysia, UNDP-FAO Regional Project RAS/86/002, pp. 10–14.

De Lange JH (1978). Shoot-tip grafting—A modified procedure: The citrus. *Subtropical Fruit Journal* 539: 13–15.

Frison EA and Taher MM (1991). FAO / IBPGR technical guidelines for the safe movement of citrus germplasm IOCV/IPGRI, Rome, Italy.

Ghini P, Marques JF, Tokunaga T, and Bueno SCS (2000). Control of *Phytophthora* sp. and economic evaluation of a solar collector of substrate disinfection. *Fytopathologia Venezolana* 13 (1): 11–14.

Gonzales CI, Mercado BG, Dimayaga BG, and Magnaye LV (1991). Recent developmental technologies towards the control of citrus greening disease in the Philippines. *In*: Rehabilitation of Citrus Industry in the Asia Pacific Region. *Proc. 6th Int. Asia Pacific Workshop on "Integrated Citrus Health Management"* (eds. Ke Chung, Shamsudin, and B Osman). Kualalumpur, Malaysia, pp. 15–20.

Hartmann HT, Kester Dale E, and Davies Fred T (1993). Plant Propagation Principles and Practices. Prentice-Hall of India Private Ltd., New Delhi.

Hong-Ji SU and Jan-Yang Chu (1984). Modified technique of citrus shoot-tip grafting and rapid propagation method to obtain citrus bud-woods free from citrus viruses and libukin organism. *Proc. Int. Soc. Citriculture.* Int. Citrus Cong. Sao Paulo, Brazil, vol. II, pp. 332–333.

Monteverde EE, Laborem G, Reyes FJ, Rulz JR, and Espinoza M (1992). Research on citrus improvement in Venezuela. *Proc.Int. Soc.Citriculture.* VII Int. Citrus Cong. Acireale, Italy, vol. II, pp. 752–755.

Mukhopadhyay S and Chaturvedi HC (2000). Citrus. *In*: Plant Tissue Culture from Research to Commercialization—A Decade of Support. Department of Biotechnology, Ministry of Science and Technology, Government of India, pp. 139–144.

Mukhopadhyay S, Rai J, Sengupta RK, and Gurung A (1997). A rapid technique for the production of nucellar seedlings of Darjeeling orange. *Proc. 5th ISCN Cong.* pp. 67–74.

Mukhopadhyay S, Rai Jaishree, Sharma BC, Gurung Anita, Sengupta PK, and Nath PS (1997) Micro-propagation of Darjeeling orange (*Citrus reticulata* Blanco) by shoot-tip grafting. *J. Hort. Sci.* 72(3): 493–499.

Murashige T and Skoog F (1962). A revised medium for rapid growth and bioassays with tobacco tissue culture. *Physiol. Plant* 15: 473–497.

Navarro L (1981). Citrus shoot-tip grafting in vitro (STG) and its applications: A review. *Proc. Int. Soc. Citriculture*, Int. Citrus Cong. Tokyo, Japan, vol. 1, pp. 452–456.

Navarro L and Suarej J (1977). Tissue culture techniques used in Spain to recover virus-free citrus plants. *Acta Horticulturae* 78: 425–435.

Navarro L, Roistacher CN, and Murashige T (1975). Improvement of shoot-tip grafting in vitro for virus-free citrus. *J. Amer. Soc.* 100 (5): 471–475.

Nicoli M (1984). La generation des agrumes en corse par la technique de microgreffege de meristems in vitro. *Fruits* 40 (2): 113–236.

Pradhan J (1993). Polyembryony in Darjeeling orange and other *Citrus* species in relation to disease resistance and mass propagation of disease-free quality seeds. Ph.D. thesis, Bidhan Chandra Krishi Vishwavidyalaya, West Bengal, India.

Roistacher CN (1991). Graft Transmissible Diseases of Citrus: Handbook for Detection and Diagnosis. IOCV/FAO, Rome, Italy, 286 pp.

Santos Filho HP, Paguio Onofre de la Rosa, Cunha Sobrinho AP, da Almir Pinto Coelho Y, et al. (1984). The Citrus Variety Improvement Program in Brazil. *Proc. Int. Soc. Citriculture*, Int. Citrus Cong. Sao Paulo, Brazil, vol. 2, pp. 329–331.

Starrantino A, Terranova G, Russo F, and Caruso A (1984). Citrus virus diseases and sanitary improvement in Italy. *Proc. Int. Soc. Citriculture*, Int. Citrus Cong. Sao Paulo, Brazil, vol. 2, pp. 329–331.

Vogel R, Bove JM and Nicoli M (1988). Le programme francis de selection sanitaire des agrumes. *Fruits* 43 (12): 709–720.

6

Postharvest Technology

The *Citrus* Industry can be distinguished into three sectors: production, harvest and utilization. Production involves field operations, harvest involves plucking of fruits form the trees and utilization involves marketing of fruits, as fresh fruits, juice or concentrate, etc. In developing countries, these sectors have little relevance although technically independent of each other. Growers in these countries are mostly small landholders and operate their orchards following traditional methods. They grow plants to produce fruits without reference to their utilization. They also harvest their products in a traditional way. But their production and harvest have no technical link. Harvesting is done manually, mostly by hired labor. Once harvested, or before the harvest, fruits are purchased by dealers/wholesalers for their marketing, either as fresh fruit or as raw materials for industries. Growers initially collect the fruits in bamboo baskets to load them in trucks. The dealers/wholesalers store the fruits in a warehouse after collecting them from different orchards, and repack them in wooden boxes to send to distant markets or industries, whichever is convenient to them. In some warehouses, fruits are treated with fungicides to minimize storage and transit spoilage. There is no production system based on potential ways of utilization of the product. As the harvest is totally a manual process, the production system remains technically unrelated to it. Generally speaking, it may be stated that the citrus industry in most developing countries lacks coherent growth and the postharvest processes remain completely neglected, resulting in substantial loss of product utilization. Growth of the orange fruit processing industries is very limited due to the poorly developed relationship between production and utilization.

In the advanced countries, the purview of the *Citrus* Industry is completely different. The growers have large landholdings and the utilization of the product is of primary concern before planning for the production. According to the type of utilization, varieties are selected including early, midseason, and late season. All subsequent operations are mechanized.

Field operations are designed according to the extent of mechanization of harvesting such as layout, planting direction, plant to plant, and row to row distances, pruning, hedging, etc. Harvesting is sometimes technically connected to processing. Production, harvest, and utilization are wholly integrated, keeping quality and spoilage loss under control. Thus postharvest technology has a different connotation in different countries and its operation varies according to the extent of industrialization in those countries. But whatever be the operational stages, there are common intervention points that may be pursued in all orchards irrespective of the economic status of the countries concerned.

INTERVENTION POINTS

The objectives of postharvest technology are: (i) plucking fruits in proper time with minimum damage to them and the harvesting machines; (ii) grading and treating fruits to preclude contamination, spoilage, and damage due to package house operation; (iii) appropriate storage to minimize weight loss, other internal chemical changes, and in situ enzymatic activities to maintain fruit quality for a long period; and (iv) to adopt measures for keeping the fruits fresh for a sufficient period of time outside the storage for their safe marketing as fresh fruits.

PLUCKING OF FRUITS: TIME AND PREHARVEST OPERATION

Plucking of fruits is done when they are at the harvest maturity stage. There are several computational and physical methods to determine this stage. It is normally specific to a variety that takes a fixed period to attain harvesting from full bloom. Accordingly, it is possible to ascertain calendar dates for plucking fruits in an orchard. Color change of fruits is also an indicator for maturity in some varieties such as kinnow oranges. But environmental conditions, particularly temperature, relative humidity and sunshine, may alter the period of ripening. To overcome this variability, several computational and physical methods are available to ascertain harvesting time. These include (i) determination of Mean Heat Units, (ii) determination of T-stage, (iii) determination of the fruit retention strength on the stalk, (iv) size and weight of fruit and its surface morphology, (v) flesh firmness, (vi) specific gravity of the juice, (vii) juice content, and (viii) total soluble solids (TSS), etc. But for all practical purposes, grower perception is most important as he/she can identify the maturity stage simply by visual inspection. Once the fruits reach maturity they are to be plucked immediately. Tree-storage can be done, however, by spraying gibberellic acid prior to color break at the rate of 10 ppm, which also causes rind softening and senescence.

In view of the market demand, delayed maturity often becomes necessary. Casagrande et al. (1999) observed that the application of Gibberellic acid (GA3) at 100 ppm delayed the maturity in 40% of the fruits for one month; the remaining fruits exhibited green flavedo and matured 68 days after the treatment. El Hammady et al. (2000) observed delay in the maturity with treatment of GA3 and calcium chloride. They observed decrease in the fruit drop by spraying 4–6% calcium chloride solution. GA3 at 10 or 20 ppm delayed the start of fruit coloring by about 10–12 days. Spraying of GA3 or calcium chloride increased endogenous GA3 levels and decreased ABA (ascorbic acid) levels. Treated fruits when stored at 5° C ± 1 and 90% relative humidity remained in good condition for about 60–75 days. TSS, acidity, TSS/acid ratio and ascorbic acid in the juice were found to be slightly affected during cold storage. Porat et al. (2001) observed that "Oroblancho" citrus fruits (*C. grandis C. paradisi*) remained green up to a period of 5 weeks when stored at 2°C after postharvest application of GA3. According to them application of GA either as preharvest spray or as a postharvest dip effectively retained the green color in cold storage.

Plucking of fruits is apparently a simple operation, but is very important from the point of view of the postharvest fruit health. If any damage, such as scratch, wound, injury, occurs at the time of plucking, it may predispose the fruits to infection that subsequently leads to spoilage. To minimize such injuries, it is often recommended to apply preharvest sprays of certain abscission chemicals to loosen the fruits from the stalks.

PREHARVEST SPRAYS OF ABSCISSION CHEMICALS

Wilson et al. (1977) reviewed the application of abscission chemicals as an aid to citrus fruit removal, and described the use of ETHREL, RELEASE, ACTI-AID, PICK-OFF, and their combinations. Applications of these chemicals have their advantages and disadvantages. They are normally selective and may not be applicable to all types of citrus varieties. Some of their merits and demerits are listed below.

ETHREL (Amchem, Amber, PA, USA): Used worldwide; applicable to different types of citrus plants, in particular, lemons, tangerines, and tangerine hybrids; normal dose 200–250 ppm. In addition to producing fruit loosening, it also enhances fruit color development.

Demerits: (i) occasional erratic performance, (ii) may cause excessive leaf-fall, (iii) not applicable with any surfactant.

El Rayes (2000) found that applying Ethrel (ethephon) at 400 ppm 15 days prior to harvest gives best results on color development and fruit ripening of Washington Navel and Amoon oranges. Its field application improved the fruit skin color, skin carotenoid content, SSC %, vitamin content, percent of fruit juice, and reduced the juice acidity.

RELEASE (Abbot, Chicago, Ill., USA): suitable for late oranges such as Valencia, causing virtually no injury to bloom, young fruit or foliage; dose 75–125 ppm to early, and midseason oranges; 175–250 ppm to late varieties of oranges.

Demerits: superficial injury of peel that often appears as distinct ring burn at the blossom-end of the fruit.

ACTI-AID (Upjohn, MI, USA): produces good loosening of early and midseason oranges; dose 10–20 ppm; suitable only for oranges intended for processing; suitable surfactants can be used along with this chemical.

Demerits: (i) erratic performance on late oranges, (ii) causes light to moderate rind pitting.

PICK-OFF (Ciba Geigy, Switzerland): similar to RELEASE in action; not applicable with a surfactant; dose 300 ppm.

Demerits: (i) produces more fruit peel injury, (ii) may cause injury to immature fruits.

When these chemicals are used in different combinations, the action of RELEASE and ACTI-AID cause superior fruit loosening in early and midseason oranges; the doses of the chemicals in combination are 50–100 ppm and 1.5 ppm respectively. This combination can be used with all surfactants.

Application of abscission chemicals, although very useful, needs appropriate expertise. For their effectiveness, most of the chemicals are to remain on the fruits at least for a few hours; sudden rains may result in complete loss of activity. Low air humidity, low temperature, and diseased or unhealthy plants often act as deterrents to the activity of several chemicals.

Hartmond et al. (2000) tested methyl jasmonate (MJ) as a potential abscission chemical to enhance mechanical harvest of Hamlin and Valencia oranges. Solution of 10 mM MJ application either as a stem wrap to individual fruit or a spray to the entire tree or the canopy resulted in significant and consistent reduction of fruit detachment force and caused fruit drop within 7–10 days. Fruit loosening is preceded by an increase in the internal ethylene concentration of fruit similar to that of other abscission compounds. Solution exceeding 10 mM MJ induced unacceptable level of leaf abscission.

PLUCKING OF FRUITS: HARVESTING

Plucking of citrus fruits or their harvesting is done manually or by specially designed machines. They are actually to be cut from the stock to avoid injury and postharvest infection. Manual operations are also aided by mechanical support. Simple manual harvesting is labor intensive and cost effective but its efficiency is very low. It is difficult to implement in big orchards. When manual handpicking is supported by mechanical means in the form of ladders or positioning equipment, the increase in efficiency is marginal but comparative operational costs become very high. To overcome this difficulty, in advanced countries such as the USA, Israel, Australia, Spain, etc., several types of harvesting machines have been designed and manufactured, of which some have proved useful. Mechanical harvesters are normally of two different types, *contact machines* and *mass removal machines* of which the latter one proved more successful.

Contact Machine

A contact machines is so designed that it can establish individual contact with loosened or normal fruits and cause them to drop to the ground or carrier. These machines employ either a spindle or a comblike mechanism to detach the fruit. They consist of a picking head, housing the detachment mechanism, and a positioning mechanism to place the head in the tree. The picking head normally consists of some array of long fingers or rotating surfaces with sufficiently large physical dimensions between them to allow passage of the foliage and young fruit, but not large enough to allow mature citrus fruit to pass. The devices include rotating rubber rollers and cones, flexible hooks, vacuum cups, and rubber spindles. It is necessary to optimize the geometry and the components of the spindle or auger picking head for efficient fruit removal. Such machines can achieve maximum fruit removal efficiency of 65% in a tree whose height does not exceed 4.0 m and which has small flexible limbs.

To operate these machines, the picking head of considerable flexibility is positioned in the fruit bearing zones of the tree; the direction of penetration is made perpendicular to the tree canopy surface to achieve maximum penetration; depth of penetration normally does not exceed one meter.

The fruit removal rate and the efficiency of such machines are low and their application in the field is not economical.

Mass Removal Machine

Mass removal machines generally employ some type of external shaking force transmitted to the fruit through the limbs or foliage. This force is generated by a mechanical vibration source attached to the trunk, limb or foliage of the tree or by oscillating air or water pulses that shake the foliage.

Limb Shakers

This device for fruit removal applies an oscillating unidirectional force to the main scaffold limbs of a tree. The force is generated by an unbalanced mass and transmitted to the limb by a boom and clamp. Two general principles are followed to generate such shaking forces: an unbalanced mass oscillated by a slide-crank mechanism and eccentrically mounted masses rotating around a shaft to develop the oscillating force. The slide-crank type of shaker can develop larger net-shaking forces; it is better suited for long strokes and low frequency shaking; the mechanism of the other shaker is simple and better suited for short strokes and high frequency shaking. A slide-crank shaker (15–17 cm stroke at 16 Hz frequency) can achieve a removal efficiency of 80 to 90%. Limb shakers may normally be positioned in two ways: manual and remote. Manual positioning has greater flexibility and is easier to operate; the remote positioning shakers have greater efficiency but their operation requires skilled workers. These shakers work well with early and midseason oranges but depress the yield of late season crops. Other disadvantages of these shakers are: (i) inherently low operational reliability, (ii) large oscillating forces delivered through long strokes at varying shaker boom angles cause fatigue, breakage and wear in the shake components and supporting mechanism, (iii) target plants have to be appropriately pruned to make the fruits visible and to increase the efficiency of the harvest, and (iv) bark damage may occur in different varieties depending on the magnitude of the shaking force.

Foliage Shakers

This fruit removal device applies oscillating forces near the outer canopy (foliage) of the tree. The forces are transmitted to the tree and/or fruit by contact. There are two types of foliage shakers: one utilizes horizontal oscillating forces to contact both fruit and foliage, and the other clamps to the foliage and applies vertical oscillating forces for fruit removal. Foliage shakers can be controlled for better precision than others. In the horizontal foliage shaker, the shaking head is inserted into the tree; the head consists of horizontal rows of flexible tines (1.0 m in length). Alternate rows of tines are oscillated in opposite directions and accomplish fruit removal by contacting both fruit and the limbs. Its efficiency has been claimed to be 70 to 85% under good operational conditions. The removal rate has been estimated as 1800 to 2700 kg ha^{-1} with a picking head of

120 cm × 120 cm. Although it is an efficient harvester, its major problem is tine breakage, inability to penetrate sufficient depth of canopy in wide varieties of tree structures. The vertical foliage shaker consists of a bank of metal tines clamping the smaller tree limbs at distances up to 60–90 cm inside the canopy to transmit the vertical shaking motion for fruit removal. Its mature fruit removal efficiency varies from 80 to 90% with a fruit removal rate about 1800 kg ha^{-1} at a shaking stroke of 20 cm at frequencies of 2.5 to 3 Hz. The major problems of this harvester are its inherently low operational reliability and breakage of metal tines use to the high vibration generated by oscillating the mechanism of clamping tines. Further its efficiency vaies in subsequent yields.

Air Shakers

Removes fruits by pulsing air blasts into the tree. Shaking forces in the direction of air movement result from the air drag on all components in the tree. Movement of the components in the opposite direction of the wind movement depends upon the spring constant of the wood. With abscission chemicals the efficiency of air shakers may be 95% with a removal rate of 18,000 kg ha^{-1}, in early and midseason oranges. However, there may be a depression in subsequent yields of 10–15%.

Water Shakers

Generally works on the same principle as air shakers to remove fruits from a tree. In this case water tends to remove the fruit by impinging on it rather than by shaking the limbs. Fruit removal is accomplished with a manually or automatically positioned water gun. Water stream from the gun sweeps the tree in an oscillatory movement. Satisfactory removal is obtained at a force 15 L s^{-1} water at a pressure of 400 kg cm^{-2} after due application of abscission chemicals. The removal efficiency achievable is 90%, the capacity of removal 18,000 kg h^{-1}. Approximately 4,000 L of water is required to remove one ton of fruit.

Trunk Shakers

Removes fruit by applying oscillating directional forces to the trunk or major scaffold limbs of a tree. The shaker head is mounted on a three-wheeled prime mover. The masses rotate about a common shaft in a shaker head that includes the clamping pads for transmitting the shaking force to the tree. At a given driving speed, the shaker head produces an omnidirectional shake that gradually changes its magnitude from minimum to maximum and vice versa. The fruit removal capacities of these shakers, using abscission chemicals, range from 9,000 to 13,000 kg h^{-1}. These shakers with abscission chemicals have little effect on subsequent yields.

Whitney et al. (2000) applied trunk catch system in commercial level and observed the increase of removal efficiency by using abscission chemicals like produfuron, metsufuron-methyl, chloridazon (CMN pyrazole), etc.

Current Generation Harvester

Harvesting of citrus remains as a problem to the growers who are involved in large-scale production of this fruit. The problem is both in operation and cost effectiveness. Because of the heterogeneity of crop types, structural differences in trees, climatic differences, etc., it is difficult to achieve uniform mechanization throughout the world. Elite growers of some countries tried to fabricate machines to meet up their own requirements, which were later field tried by their citriculturists. Some of the mechanical harvesters constructed in various countries have been described above but each has its own limitations. Today efforts are underway in different countries notably Japan, USA, France, and Italy, to apply modern technology to construct a Robotic Harvester. France and Spain initiated a joint project in 1987, CITRUS-ROBOT, to develop a citrus harvesting robot (Juste et al., 1992). The concept applied in this study is spherical coordinates and telescopic movement with gradual modification of the various stages, the size, configuration, and motorization of the arm and grasping element. Required agronomic conditions are first studied to adapt a robot for work. Variety, tree size, pruning method, orientation, fruit distribution, foliage density, etc. are related to fruit visibility and detection. Accordingly, suitable agro-techniques, vision systems, manipulator design, and grasping principles are being investigated to develop suitable robots for citrus fruit harvesting.

GRADING AT THE PACKINGHOUSE

Fruits may be plucked manually or by machines. Hand-plucked fruits are put immediately in containers carried by the pluckers on their back. Machine-aided manually plucked fruits are also put in the containers immediately after plucking. In case of mechanical harvesting, fruits are either dropped on the ground (ploughed before operation) or collected in an attached padded catch frame. A fruit rake and pick-up machine are also used to collect the fruits from the fields. These fruits are taken to the packaging houses for further operations. Plucked fruits from the fields are normally transported to the packinghouses by trucks. Trucks may be loaded manually or mechanically. In the mechanical operation, catch frame systems contain pallet bins with tractor forklift, plastic, and wire tubs, and loader boom. The bins are directly carried to the trucks by manipulating the tractor forklift. Tub or basket fruit containers are unloaded with the loader boom mounted on the high lift truck and the empty ones are rolled or tumbled by hand to the fruit collector bin of the catch frame. Once the fruits reach the packinghouses, they are graded as fast as possible for further processing. Grading is normally done to separate out off-types (color, size, shape, weight, etc.), damaged, and infected/infested fruits. This can be done manually or mechanically. In a mechanical system, unloading, grading, and subsequent processing are integrated, and operated electrically.

Grading Machines

Different types of machines are available for commercial grading purposes. A typical machine is Pennwalt Corporation's Electroscan™ (Kaplan et al., 1984). The Electroscan consists of a conveying system approximately 7.0 m long, and 3.0 m wide, about the size of a grading table, and computer console. Fruits are singularized into nine lanes, each capable of sorting 7.5 fruits s^{-1}, classified, and distributed onto two takeoff belts or pass directly through the unit with 80% load efficiency. This equipment can scan 27,000 fruits h^{-1}.

The simulator is designed to position the fruit between two rollers rotating by a belt and providing positive angulation. Fruits with their stemblossom remaining perpendicular to their forward motion continue to spin while under the camera. The camera consists of photodiodes, specialized filters, and color detector. The photodiodes convert incident light into a voltage. The reflection of the surface of the fruit is filtered to enhance the defect. Through a unique optical system, the fruit is illuminated from four sides. The video processors control the camera's scanning. There is one video processor per lane of fruit sorting. Analog voltages from the photodiodes are converted to digital signals for computer processing. These numbers are then reduced to a set of numerical values that are proportional to blemish, color, size, variegation, etc. The values of each piece of fruit are transferred to the master processor computer and are normalized on a 0 to 100 scale for ease of the operator. These values are compared to the operator-determined breaks and the fruits categorized. The computer uses a lookup table to determine the destination of the fruit in each category. If the fruit is to be diverted from the simulator onto a takeout belt, the master processor activates a solenoid that delivers air pressure to the diverter's pneumatic cylinder. The duration of the diversion stroke is less than 50 ms.

The operator interfaces with the system through a display terminal connected to the master processor computer. Operational status of the system components is displayed as a result of the selfdiagnostic tests. The operator unit runs a simple calibration procedure that scales the video data so that all the lanes grade equally. There is a choice of algorithms for calculating blemish and color. The blemish and color parameters are combined with size and variegation to properly classify the fruit.

Another automatic video grading machine is Video Grading™ (Blandini et al., 1992). It is controlled by a microcomputer through an artificial vision system. This machine is built in a modular way and can be fitted in any ordinary processing lines since it is perfectly interchangeable with traditional sorting lines. The control unit is composed of two Motorola CPUs MC 68020 boards each with 01 M byte of dual ported RAM, and two serial interfaces, a 1/10 card with 8 opto-isolated inputs, 8 opto-isolated outputs, 8 relay outputs, and a disk controller with 20 M bytes ST-506 hard disk and 01 M byte SA-450 floppy disk.

One CPU, the master, is linked through its serial ports to a VT-100 terminal (for technical operation and debugging) and to a custom control panel with mini-pointer that prints a ticket with all the data about the orange processed. The control unit mounted into a rack together with the vision system and power supply, can control up to four selection lines. Each of these units can process up to 86,400 fruits h^{-1} or 10,000 kg h^{-1} (with a 120 g average mass per orange). There is also scope to add more units modularly to meet the user's needs. The vision system is composed of the hardware for image acquisition, a Pulmix CCIR, monochrome CCD video camera, and a Defect Analysis Board (DAB). It processes the pixels coming from the A/D converter on fly during acquisition. The results are a set of counters that are further processed by the CPU at frame-rate speed. The pixels are binarized through a two-stage thresholding process: the first stage is used to separate the orange from the background; the second is adaptive and good at locating small defects. The vision chamber is a sheet-metal box painted white on the inside. Diffuse illumination is obtained using neon lamps driven by a "Quicktronic" to get high frequency lights and precludes flickering. The camera is mounted on top of the chamber to view two oranges in the same scene or to get two different views of the same orange at different times. Inside the vision chamber, there are two mirrors placed close to the fruit to see the lateral surface of each orange.

To organize the working session with Video Grading, orange boxes are unloaded in a conveyor belt with rollers. Fruits are rotated 180 in one roller conveyor step, allowing inspection in two different views of the same orange with a large percentage of inspected surface. The actual speed is set to 6.26 fruit s^{-1} corresponding to four frame periods (160 seconds).

Geensill and Newman (2001) advocated the use of a *Near-Infrared* (NIR) spectrometer with differing dispersion elements for in-line fruit sorting. They assessed the practical application by recording the spectra of transmitted radiation from the whole immature and mature limes over a wavelength range of 650–1,050 nm.

POSTGRADING OPERATION

If the fruits can be properly graded, quality fruits of proper size, shape, weight, and color can be separated out for storage preceded by degreening operation using suitable chemicals such as ethylene. To make such operations more effective and acceptable, Miller et al. (2000) suggested the application of "sensor technology" for real-time process control of ethylene as its application at >10 ppm enhances decay. They could monitor the level of ethylene by chemical luminescence (CL) calibrated by gas chromatography equipped with a flame ionization detector (GC-FID) and adapted at different locations. According to the authors such monitoring is necessary to assess the quality by controlling the ethylene level as well as the standard parameters of temperature, humidity and fresh air exchange in the degreening process.

In the past, cold temperature storage was the only option. But storing fruits in cold (2° to 4° C) cannot control their decay. On the contrary, the extent of decay may increase with the extent of the storage period. The efficacy of the cold storage on the preservation of citrus fruits largely depends upon the variety, season and fruit maturity stage, canopy depth, and harvesting date. Grapefruits are more sensitive to chilling injury (CI) than oranges and lemons. Navel oranges grown on different rootstocks differed in their CI sensitivity (Arpaia, 1994). Maturity age, determined by harvesting period, greatly influenced CI susceptibility. CI has been found to be highest in lemons, oranges and grapefruits in early-, moderate in mid-, and negligible in late-season picked fruits (Schirra et al., 1998, 2000). Fruits in storage usually deteriorate for various reasons, among which loss of weight and decay due to fungal infection are very important. Initially, it was thought that both these deterrent factors might be minimized by cold temperature treatment. It is true that cold temperature may minimize weight loss by reducing the rate of respiration, inactivating the in situ enzymatic activities, and other chemical reactions. But fungi are differentially sensitive to low temperature and CI.

POSTHARVEST INFECTION BY FUNGI

A large number of fungi can infect citrus fruits on storage. Tuset (1984) recorded 15 species and their relative incidence in storage and marketing causing rots (Table 6.1). His observations revealed the significance of *Penicillium* and *Alternaria* species in causing rots to stored fruits. It was

Table 6.1: Fungi causing postharvest fruit rots under Spanish conditions

Fungi		Incidence of rots (%)[a]	
		During storage[b]	During marketing
1.	*Penicillium digitatum* Sacc.	30–55	70
2.	*Penicillium italicum* Wehmer	17–37	10–25
3.	*Alternaria citri* Ell.	10–45	0–20
4.	*Botrytis cineria* Pers.	02–15	0–10
5.	*Colletotrichum gloesporoides* Penz.	0–07	0–03
6.	*Geotrichum candidum* Link.	0–05	0–03
7.	*Rhizopus nigricans* Ehrenh.	0–04	0–03
8.	*Cladosporium herbarum* Link.	0–02	0–01
9.	*Alternaria alternata*	0–01	0–0.8
10.	*Phytophthora citrophthora* (Sm. and Sm.) Leonia	0–03	0–05
11.	*Trichothecium roseum* (Bull.) L.K	0–0.8	0–01
12.	*Fusarium oxysporum* Sehl.		
13.	*Fusarium solani* (Mart.) App. and Wr.	0–1.5	–
14.	*Diplodia mutila* (Frias) Mont.	0–0.1	–
15.	*Phomopsis citri*	0–0.8	–

[a]oranges (Navel, Salustiana, Valencia late) and mandarins
[b]conventional atmosphere, 90% relative humidity, after two months at 2–4° C

further shown that these fungal infections occur even after storing fruits at 2–4° C for two months.

Tuset and Marti (1988) observed that storing fungicide-treated fruits in cold failed to control *Penicillium* rot. On the contrary, when fungicide treatment in packinghouses against *Penicillium* is implemented for several years, it created conditions for other minor fungi to flourish. Ben-Yehoshua et al. (1987) made similar observations. They developed a curing technique to reduce the decay by *Penicillium* based on its temperature sensitivity. When the fruits were cured at 35° C and 98% helative humidity for 70 h (not later than 48 h after harvest), *Penicillium* was eliminated. But the authors also found that the conditions that cure *Penicillium,* favor the growth of *Aspergillus niger* Van Tieghem and *Rhizopus oryzae* Went and Greer, which cause serious decay of citrus fruits. Tuset et al. (1992) further studied the relationship between the temperature and those two fungi. Both these fungi are more or less tolerant to high temperature (Table 6.2). Certain postharvest practices some times induce the infection by *Colletotrichum gloesporoides,* causing anthracnose disease in stored fruits,

Table 6.2: Temperature relationship of *Aspergillus niger* and *Rhizopus oryzae*

Temperature condition	*A. niger*	*R. oryzae*
Optimum	35° C	32° C
Favorable growth	40° C	37° C
Growth inhibition	45° C, 10° C	50° C, 05° C
Slow growth	15° C 45° C,	10° C

which is otherwise a minor pathogen. This disease is a serious one in early season "Robinson" tangerines and "Star Ruby" grapefruit. The fungus directly penetrates the healthy rind when the fruits are degreened with ethylene. Degreening also enhances spore germination and formation of appressoria. Anthracnose infection is enhanced if the degreened fruits are waxed with resin solution water wax (Brown, 1992).

Roy (2000) observed the scope of postharvest application of hot water in various citrus fruit species before packinghouse treatment to minimize decay due to fungi. A 2-minute dip in water at 53° or 56° C markedly reduced the decay caused by *P. digitatum* and *P. italicum* in kumquats. But treatment at a higher temperature increased the decay. In Marsh grapefruits a combined hot water (52° C) + growth regulator (50 ppm Progibb + 200 ppm 2–4 D) treatment reduced the decay. In Valencia oranges, 400 ppm imizalil heated to 55 C shows considerable control of the decay. Brown and Chambers (2000) evaluated the efficacy of polyhexamethylene biguanide (PHMB) for the control of postharvest diseases. Aqueous application of PHMB (1,000–4,000 ppm) to citrus fruits (Sunburst tangerine, Orlando tangelo, Ambersweet and Hamlin oranges) provided significant control of stem end rot caused by *Diplodia natalensis,* sour rot caused by *Geotrichum citri-aurantii* and green mold caused by *Penicillium digitatum.* Control of stem end rot in Amberswet, Hamlin oranges and mandarins was not as effective as the control observed with thiabendazole (TBZ), 700 and 1, 000 ppm TBZ before and after degreening respectively. Sour stem rot control in tangelos by PHMB at 2,000 or 3,000 or 4,000 ppm was significantly better than that found with sodium ortho-phenyl phosphate (SOPP) at 20,000 ppm. Efficacy against green mold in Hamlin oranges was similar to the efficacy of 1,000 ppm TBZ and 1,000 ppm imizalil. PHMB at 250 or 500 ppm effectively reduced the propagule viability of *G. citri-aurantii.*

PACKINGHOUSE TREATMENTS AND STORAGE

Packinghouses very often have to deal with the variegation of fruits. Sometimes fruits are harvested when they reach internal maturity with no change in external color. In packinghouses they are *degreened* by chemical treatments. *Ethephon* (active principle of Ethrel) applied at the rate of 4,000 ppm under warm conditions (Temperature 25° C or above) can break the green color to make the fruits yellow. Degreening also occurs by exposing the fruits to *ethylene gas* (5–50 μl^{-1} for 90 h depending on the type of the fruit), but this gas induces infection by several fungi and also enhances spore germination.

Attempts have been made from time to time to reduce the rate of respiration by applying antitranspirant chemical, such as *Vapor guard*; exposure to carbon-dioxide (40% for 24 h at 20° C); waxing, etc. Among these, waxing has been widely used along with temperature conditioning to reduce fungal infection. There is also scope for preharvest hormonal treatment and UV irradiation as postharvest treatment since such treatments induce resistance in *Citrus* sp. and hybrids by increasing the concentration of flavonoids (Del Rio and Ortuno, 2003).

POSTHARVEST OPERATIONS: WAXING

Wax is available in two forms—*solvent wax* that contains 10% coumarone-indene resin and *water wax*, which contains carnauba, polyethylene, paraffin, lac, colophony, and coumarone-indene. Cuquerella et al. (1981) studied the physiological effects of different wax treatments on *Citrus* fruits in cold storage. They used solvent wax and three different types of water wax differing in composition as follows:

04% polyethylene, 21% paraffin, 04% lac, and 04% colophony; (ii) 06% polyethylene, 04% lac and 04% colophony and (iii) 06% carnauba, 04% lac, and 04% colophony. Solvent wax is avoided because its production involves loss of nonrenewable fuel and leads to polluted effluents. Hence water wax containing a high quantity of polyethylene is used in the citrus industry. But the effect of waxing is not the same in different citrus fruits. It can reduce the shrinkage of Valencia and Naval oranges but not Clementine mandarins. Consumers by and large do not like wax-treated fruits. In waxed fruits, internal carbon dioxide concentration increases and oxygen concentration decreases; this modification is related to ethylene production, volatile build-up, and off-flavor development.

Temperature conditioning has been described along with the packaging of different citrus fruits. Chen et al. (2000) found that fruit quality remains unaffected when coated with "Rinrei's wax" before plastic bag packaging and storiage at 1° C gives for 12 weeks. The concentration of wax applied depends on the species. The authors found the best concentrations for wax coating for fruits of Shaddock, Ponkan mandarin,

Liucheng orange, Tanaka tangor, grapefruit, and Haili tangor were 50, 40, 60, 30, 70, 100% respectively.

The exact mode of action by which waxing reduces CI is not clearly understood. One possible mechanism is reduction in rate of gas exchange through the peel, modifying the fruit's internal atmospheric composition and resulting in increased carbon dioxide and decreased oxygen levels (Hagenmaier and Baker 1993; Petracek et al., 1999).

HEAT TREATMENT

Porat et al. (2000) developed a *hot water brushing technique* (HWB) to control green mold decay of red grapefruit and also to control chilling injury (CI) during cold storage. This technique involves drenching the fruits with hot water on the fruits as they move along a belt of brush rollers. A *20*-s HWB treatment at 59° or 62° C reduced the decay and CI index. It improves general appearance but did not influence fruit weight loss, % TSS (total soluble solids) in the juice, juice acidity or fruit color. Rodov et al. (2000) cured by treating Orobanco fruits at 36° C for 72 h followed by hot water dip at 52° C for 2 min or hot drench brushing at 52, 56 or 60° C for 10 s. The cured fruits were then given standard packing treatments including waxing with the addition of thiobendazole (TBZ) and 2,4-D isopropyl ester followed by low temperature treatments (particularly for exports). Treated fruits were stored at 1° C for 2 weeks (simulated low temperature quarantine treatment) followed by 12–13 weeks (simulated sea transport), and additional one week at 20° C (simulated retail shelf-life period).

FILM PACKING

Introduction of polyethylene films in citrus storage has opened up new vistas in the storage of fresh fruits. Film wrapping usually has a beneficial effect on the preservation of cold-stored fruits. Large-scale commercial use of polyethylene bags (nonsealed Unipack low density film bags) storing 100,000 metric tons of Hassaku and Amantsu has been practiced in Japan since 1970s. Normally a suspension of High Density Polyethylene (HDPE) "Tag" gives better results in delaying fruit deterioration than wax materials. Tag's action relates to inhibition of the transpiration process without affecting fruit flavor. Low Density Polyethylene (LDPE) bags are also used for storing citrus fruits but augment decay through water condensation in the bags.

Martinez-Javega et al. (1981) in Spain studied utilization of polyvinyl chloride sealed packaging of *Citrus* fruits. They used polyvinyl chloride extensile film (0.016 mm) to seal Washington Navel and Valencia oranges and found the following advantages of unwaxed but sealed fruits: reduction in respiration rate, prevention of chilling injury, reduction in decay, reduction in fruit, carbon dioxide concentration, reduction in ethylene gas production, delay in softening, and no increase in juice volatile content. Waxed fruits, on the contrary, had an off-flavor.

Wrapping fruits in HDPE gives good results but sealing the fruit in this film gives even better results. For this purpose, fruits are seal-packed individually in HDPE (10 μ film). This film can be obtained from Mitsui Petrochemicals, Japan and Hoechst, Switzerland. Such packaging induces the following effects: 1) doubles the storage life of fruit, retaining freshness and flavor without shrinkage; 2) five-fold reduction in weight loss; 3) inhibition of chilling injury; 4) enhancement of degreening; 5) production of desirable color substituting the commercial degreening by ethylene; and 6) acceleration or inhibition of ripening by reinforcing the package with suitable chemicals. 7) This seal packing further inhibits growth of molds but enhances the growth of stem-end rots for fungicidal treatment needs to be done before packaging. 8) HDPE packaging can sometimes replace cooling. Packaged grapefruit and Valencia oranges at 20° C perform better than unpacked fruit stored at 10° C and 02° C respectively. Fruit sealed in other films such as LDPE or the thinnest polypropylene film (20 μ) usually have an off-flavor.

Ben-Yehoshaua et al. (1981) continued the studies on HDPE packaging of *Citrus*, concentrating on its effectiveness for lemons. The fruits were conventionally washed, disinfected with sodium orthophenylphenate (SOPP), rinsed, dried, and waxed (containing polyethylene and 4,000 ppm thiabendazole), then seal packed with HDPE film (specific density 0.995). The authors observed marked delay in deterioration. According to them, HDPE sealed packing effects are related to the water-saturated microatmosphere in the sealed enclosure around the fruit; the atmosphere elevates the water-stress that exists in harvested fruit; in sealed fruits, symptoms of water stress were prevented and firmness was maintained.

OTHER TREATMENTS

Castro-Lopez et al. (1981) also worked on the storage of lemons in Cuba, but did not use polyethylene wrapper. According to them, picking time significantly influences oleocellosis due to damage of oil glands of fruits. Fruits were disinfected by SOPP and treated with fungicides, benzimidazoles, benomyl, and thiabenzadole (TBZ). Then they were cured by keeping them at 15° C (relative humidity 95%) for 48 h, and stored at 10° C for 42 days without CI. The efficacy of minimum concentration of thiabendazole (TBZ), imizalil and sodium ortho-phenylphenate (SOPP) was reexamined by Cohen et al. (1992). They found that at 250 ppm, the efficacy of TBZ and imazalil did not change. But in grapefruit, reduced TBZ concentration increased CI sensitivity. Heat treatment with 250 ppm TBZ at 48° C increased fungicide activity but reduce CI and decay in long-term cold stored grapefruit. TBZ proved more effective at low concentration (200 mg L^{-1}) in hot water dips (53° C) (Schirra et al., 2000). Replacing SOPP at 5,000 ppm with OPP orthophenyl phenol at 2,000 ppm or 400 ppm at 45° C also resulted in effective decay control. When calcium was applied with the fungicide, the fruits acquired resistance to diseases. Chalutz et al. (1992) observed the biocontrol of postharvest diseases. Yeasts (*Pichia guilliermondii* and *Hanseniaspora uvuarum*) at 10^4 to 10^7 cfu ml^{-1} inhibited disease development in the presence of 2% (w/v) different aqueous calcium salts. But this remedy has yet to be commercialized.

Although there are various recommendations regarding the fungicidal treatments depending upon the type of *Citrus* fruits, the use of polyethylene wrapper or sealed-packages has dominated international research on storage of *Citrus* fruits, with sharp differences of opinion on the type of polyethylene to be used—HDPE or LDPE. Use of HDPE sealed package originated in Israel and was also advocated in Australia. Tugwell and Gillespie (1981) recommended a package of practices for storage of Washington Navel oranges, Ellendale mandarins, Marsh grapefruit, and Eureka lemons using HDPE wrappers. Fruits after harvest are dipped in Panoctine® (guazatine), flood-treated with Benlate® (Benomyl), and waxed with a water-based wax containing 500 ppm 2,4-D. Each fruit is individually wrapped in HDPE using a machine. Grapefruit and lemon are stored at the ambient temperature for several months. Early Washington Navels can be stored for three months but not late ones. Fungicide treatments protect the fruits from blue and green molds but not stem-end rots, particularly in overmatured or late-stored Ellendale mandarins and Valencia oranges.

The USA and Spain advocate the use of LDPE. Kawada et al. (1981) in the USA tested the unipack low-density polyethylene film for storing citrus fruits. According to them, wrapping individual fruit in unipack not only keeps it fresh in terms of minimizing weight loss, softening and changes of peel color and gloss, but it also reduces fruit deformation, and chilling injury. The principal benefit of unipack on citrus fruits is the reduction in water loss. It could replace waxing. Film wrapping loose skin type citrus fruits like mandarins is not beneficial, because it causes peel-puffing and thus increases the decay. When adequate fungicides are used, decay is often reduced because there is no water condensation between tightly wrapped film and the peel. The unipack prevents decay problems such as cross infection, so called "soilage", and the wetting of fiberboard containers. They stored Grapefruit, Navel oranges, Nova tangelos and Robinson tangerines wrapped with LDPE 4 weeks at 21° C with two diphenyl pads per carton.

D'Aquino et al. (2001) dipped Salustiana orange in 500 mg L^{-1} imazalil emulsion and then wrapped them in plastic film (MR type of medium permeability to gases and water vapor). Treated fruits were stored at ambient temperature (18–20° C) and 60–65% relative humidity for 6 weeks and no change in their initial chemical and aesthetic characteristics was found.

D'Aquino et al. (2001a) further observed the usefulness of medium density film as a packing material to maintain the quality and physiological conditions of stored fruits. They tested three different types of plastic films (least permeability, medium permeability, and perforated) and found the superiority of films of medium permeability to inhibit aging and losses of overall quality during storage at different temperatures. D'Anna et al. (2001) used continuous refrigeration systems in containers and sensors to maintain and monitor temperature inside the containers. Prerefrigeration of fruits on farm, use of only one species of fruits and packaging material within one load of produce, proper positioning of sensors inside the fruit loads, the operation of sensors during transport, etc. are prerequisites for the success of such storage.

Palou et al. (2001) treated fruits with sodium carbonate (2–4%) or sodium bicarbonate (1–4%) at room temperature for 150 s, or at 45° C for the same period and hot water at 45° C for 60 or 50 s and stored them at 3° C for 3 weeks for effective control of blue mold.

POSTHARVEST STORAGE: LIMITATIONS AND NEW APPROACHES FOR STABILITY

The major problems in postharvest storage of *Citrus* fruits are deterioration in quality, loss of weight and decay due to biotic and abiotic causes. Methods to minimize these problems have already been discussed. But the problem that persists is the chilling injury (CI). Low temperature treatment is considered the most effective method of maintaining fruit quality, retarding respiration and ripening, and minimizing decay. It also decreases the accumulation of ethylene around nonclimacteric produce during marketing, resulting in an increase in their postharvest life (Wills et al., 1999). But this treatment has limitations also. Firstly, what should be the temperature, how long can the fruits be stored at that temperature, how to integrate different operations—storage, packaging, loading and transport—to keep the temperature steady during all the operations from the storage to the market. The total operation may involve several days to weeks.

Each type of fruit has its own optimum low temperature for its storage. For exmple, oranges and mandarins are sensitive to temperature below 5–6° C whereas grapefruits and lemons are sensitive to temperature below 11–13 C. If oranges and mandarins are stored below 5–6° C, they will be vulnerable to CI. Grapefruits and lemons, on the other hand, may be vulnerable to CI below 11–13° C. The storing temperature specific to a particular fruit can be uniformly maintained during all the operations by coupling the freezing system with a heating system using proper thermostat and sensor devises. Any deviation of the cold-temperature-chain during different operations or confinement of the commodity for a long period in the cold environment or even sudden exposure of the cold-treated fruits to a high temperature often induces CI. There is also the question regarding the threshold level of ethylene during storage. The threshold level of ethylene action on nonclimacteric produce is well below 0.005 ppm, in contrast to the commomnly considered level of 0.1 ppm (Ladaniya, 2003). Thus the composition of the effective cold chain for storage and transit to avoid CI and fungal infection still remains an unsolved problem. Storage and transit of the fruits at a temperature above the sensitivity level favor fungal infection. Storage below the optimum low temperature induces CI. Fruits may exhibit cumulative time-temperature of chilling temperature on carbon dioxide evolution, which may cause qualitative changes in the fruits. It may also increase production of ethylene and volatile components of fruits when they are transferred to normal temperature. Further cold treatment again is obligatory as quarantine treatment (14 days at 1.5° C) for export purposes, tempting risk of CI.

Therefore the only option left is to enhance the cold tolerance of fruits or development of genetically improved cultivars with enhanced cold resistance. This is possible if the problem is looked into by appropiate biotechnological methods.

BIOTECHNOLOGY OF COLD TOLERANCE

Porat (2003) recently reviewed the current knowledge on the biotechnology of cold tolerance. Cai et al. (1995) islated two COR (cold regulated) genes (COR 11, COR 19) from leaves of cold-tolerant citrus relative *Poncirus trifoliata*. These genes encode a group of small hydrophilic proteins that are similar to late embryogenesis (LEA) proteins. COR genes are specifically activated by exposure to low temperature. But these genes persist in *P. trifoliata* but disappear from chilling sensitive pomelo plants when introduced into it. Extensive research is currently in progress in various countries to isolate cold resistance genes from fruits. Sanchez-Ballesta et al. (1996) in Spain isolated several cDNAs from mandarins

(Fortune) activated by a prestorage curing treatment. Several grapefruit cold-tolerant genes have been isolated in Florida. Several genes activated by prestorage hot-water rinse and brushing treatment have been isolated in Israel (Porat, 2003).

Postharvest operations form a very strong component in storage and marketing of all types of citrus fruits. But these operations start from the preharvest stages of the crops. Major problems in postharvest operations are diseases caused by various fungi, among which *Penicillium* species are common. All fungi produce innumerable spores that remain in the environment and infect fruits under favorable conditions. Although cold storage of fruits is one means of reducing decay due to diseases, it often aggravates the fungal problem and incorporates another, namely, chilling injury. Fungicidal treatment is another means of controlling diseases but prolonged usage often induces resistance to pathogens. In such a complicated situation the primary option would be prevention, involving grower and packinghouses. *Grower must be made aware of the need for maintaining a high level of sanitation in fruit groves. They must also ensure that no fruit is damaged at the time of harvest. Should a fruit be damaged, it is to be removed immediately and not allowed to become a source of inoculum. Packinghouses for their part must maintain a high level of sanitation in their premises and storage facilities by periodic fumigation or sprays.* It is desirable to monitor the abundance of fungal spores in their houses and storages regularly and to schedule spraying/ fumigation accordingly. Packinghouses should ensure that the time gap between harvest and treatments is as minimal as possible and treatment modifications done as and when necessary. *The ultimate option, however, is isolation of cold-resistant and infection resistant genes and production of transgenic plants following suitable genetic engineering strategies.*

REFERENCES

Arpaia ML (1994). Preharvest factors influencing postharvet quality of tropical and subtropical fruit. *Hort. Sci.* 29: 982–985.

Ben-Yehoshua S, Shapiro B, and Even-Chen Z (1981). Mode of action of individual seal-packaging in high density polyethylene (HDPE) film in delaying deterioration of lemons and bell pepper fruits. *Proc. Int. Soc. Citriculture* Int. Citrus Cong. Tokyo, Japan, vol. 2, pp. 718–721.

Ben-Yehoshua S, Barak E, and Shapro B (1987) Postharvest curing at high temperatures reduces decay of individually sealed lemons, pumelos, and other citrus fruits. *J. Amer. Soc. Hort. Sci* 112 (4): 658–661.

Brown GE (1992). Factors affecting the occurrence of anthracnose on Florida Citrus Fruit. *Proc. Int. Soc. Citriculture,* VII Int. Citrus Cong. Acireale, Italy, vol. 3, pp. 1044–1048.

Brown GE and Chambers N (2000). Evaluation of polyhexamethylene biguanide for control of postharvest diseases of Florida citrus. *Proc. Florida State Hort. Soc.* 112: 118–121.

Cai O, Moore GA, and Guy CL (1995). An unusual group 2 LEA gene family in citrus responsive to low temperature. *Plant Mol. Biol.* 29: 11–23.
Casagrande JG, Bianch VJ, Strelow EJ, and Facinello JC (1999). GA3 influence on maturation of tanger 'Taquari'. *Revista Cientifica Rural* 4 (2): 34–37.
Castro-Lopez T., Izquierdo I, and Perez L (1981). Response of lemons to various harvesting and postharvest treatments. *Proc. Int. Soc. Citriculture,* Int. Citrus Cong. Tokyo, Japan, vol. 2, pp. 737–739.
Chalutz E, Droby S, Cohen L, Weiss B, Daus A, Wilson C, and Wisniewski M (1992). Calcium enhanced bio-control activity of two yeast antagonists of citrus postharvest diseases. *Proc. Int. Soc. Citriculture,* VII Int. Citrus Cong. Acireale, Italy, Vol. 3, pp. 1066–1069.
Chalutz E, Droby S, Cohen L, Weiss B, Daus A, Wilson C, and Wisniewski M (1992). Calcium Enhanced Bio-control Activity of Two Yeast Antagonists of Citrus Postharvest Diseases. *Proc. Int. Soc. Citriculture* (Volume 3): 1066–1069.
Chen-RuYin, Meeijui TSGi, Liu Ming Sai, Chen RY, Tsai MJ, and Liu MS (2000). Effect of plastic bag packing and fruit coatings in the stoage quality of citrus fruits and melons. *J. Chinese Soc. for Hortic.* 46 (1): 35–44.
Cohen E, Chalutz E, and Shalom Y (1992). Reduced chemical treatments for post harvest control of citrus fruit decay. *Proc. Int. Soc. Citriculture,* VII Int. Citrus Cong. Acireale, Italy, vol. 3, pp. 1064–1065.
Cuquerella J, Martinez-Javega JM, and Jimenez-Cuesta (1981). Some physiological effects of different wax treatments on Spanish citrus fruits during cold storage. *Proc. Int. Soc. Citriculture,* Int. Citrus Cong. Tokyo, Japan, vol. 2, pp. 734–736.
D'Anna K, Amico C, Lanza G, and Menzoa A (2001). The cold treatment of citrus fruits. *Informatore Agrario* 57 (19): 61–64.
D'Aquino S, Angioni, M, Schirra, and Agabbio (2001a). Quality and post physiological changes of film packed ' Malvasio' mandarins during long-term storage. *Lebensmittal Wissenchaftund Technologie* 34 (4): 206–214.
D'Aquino S, Molinu MG, Piga A, and Agabbio M (2001). Influence of film wrapping on quality maintenance of "Salustiana" oranges under shelf-life conditions. *Italian J. Food Sci.* 13 (1): 67–100.
Del Rio JA and Ortuno A (2003). Biosynthesis of flavonoids in Citrus and its involvement in the antifungal defense mechanism. *In*: Crop Management and Postharvest Handling of Horticultural Products 4: Diseases and Disorders of Fruits and Vegetables (eds. R Dris et al.). Science Publ. Inc. Enfield, NH, USA, pp. 1–32.
El Hammady AM, Abdel Hamid M, Saleh M and Saleh A (2000). Effect of gibberellic acid and calcium chloride treatment delaying maturity, quality and storability of "Balady" mandarin fruits. *Arab Univ. J. Agric. Sci.* 8(3): 755–766.
El Royes DA (2000). Enhancement of color development and fruit ripening of "Washington Navel" and "Amoon" oranges by Ethrel pre-harvest application. *Assiut J. Agric. Sci.* 31 (2): 71–87.
Greensill CV and Newman DS (2001). An experimental comparison of simple NIR spectrometers for fruit grading applications. *Appl. Eng. in Agric.* 171 (1): 69–76.
Hagenmaier ED and Baker RA (1993). Reduction of gas exchange of citrus fruits by wax-coating. *J. Agric. Food Chem.* 41: 283–287.
Hartmond U, Yuan Rongcai, Burns JK, Grant A, Kender WJ, and Yuan RC (2000). Citrus fruit abscission induced by methyl jasmonate. *J. Amer Soc. Hortic. Sci.* 125 (5): 547–552.
Juste F, Formes F, and Sevila F (1992). An approach to robotic harvesting of citrus in Spain. *Proc. Int. Soc. Citriculture,* VII Int. Citrus Cong. Acireale, Italy, vol. 3, pp. 1014–1018.

Kaplan HJ, Dave BA, and Petrie JF (1984). Tolerance of Citrus Pathogens to Current Packinghouse Treatment. *Proc. Int. Soc. Citriculture,* Int. Citrus Cong. Sao Paulo, Brazil, vol. 2, pp. 788–790.

Kawada K, Wardowask WF, Grierson W, Albrigo LG, and Hale PW (1981). "Unipack": Individually wrapped storage of citrus fruits. *Proc. Int. Soc. Citriculture,* Int. Citrus Cong. Tokyo, Japan, vol. 2, pp. 725.

Ladaniya MS (2003). Citrus: Postharvest cold chain. *In*: Postharvest handling of Horticultural Products 2: Fruits and Vegetables (eds. R Dris et al.) Science Publ. Inc. Enfield, NH USA, pp. 239–276.

Martinez-Javega JM, Jimenez-Cuesta, and Cuquerella J (1981). Utilization of polyvinyl chloride (PVC) film for individual seal packing of citrus fruit. *Proc. Int. Soc. Citriculture,* Int. Citrus Cong. Tokyo, Japan, vol. 2, pp. 722–724.

Miller WR, Ismail MA, and Richard R (2000). Review of sensor technologies for real time process. ASAE Ann. Int. Mtg., Wisconsin, USA, 9–12 July, pp.1–10.

Miller WR, McDonald RE, and Chaperro J (2000). Tolerance of selected orange and mandarin hybrid fruit to low dose radiation for quarantine purposes. *Hort. Sci.* 35 (7): 1288–1291.

Palou L, Smilanik JL, Usali J, and Vinas I (2001). Control of postharvest blue and green molds of oranges by hot water, sodium carbonate and sodium bicarbonate. *Plant Diseases* 85 (4): 371–376.

Petracek PD, Hagenmaier RD, and Dou H (1999). Waxing effects on citrus fruit physiology. *In*: Advances in Postharvest Diseases and Disorders of Citrus Fruit (ed. M Schirra). Research Signpost Publisher, Trivandrum, India, pp. 71–92.

Porat R (2003). Reduction of chilling injury disorders in citrus fruits *In*: Crop Management and Postharvest Handling of Horticultural Products. 4: Diseases and Disorders of Fruits and Vegetables (eds. R Dris et al.) Science Publ. Inc., Enfield, NH USA, pp. 1–11.

Porat R, Feng KuQuiao, Huberman M, Galili D, Goren N, Goldschmitt EE, and Feng KQ (2001). Gibberellic acid shows postharvest de-greening of 'Oroblanco' citrus fruits. *Hort. Sci.* 36 (5): 937–940.

Roy SK (2000). Postharvest application of hot water treatment in citrus fruits: the road from the laboratory to the packinghouse. *In*: Proc. Int. Hortic. Cong. Part 8. Quality of Horticultural Products: Storage and Processing of Underutilized Fruits of the Tropics (eds. S Ben-Yehoshua, J Petez, V Rodof, B Nafussi, O Yekutieli, A Wiseblum, R Regev, M Herragods, B Nicolai, and A de Jager). *Acta Horticulturae* 518: 19–28.

Rodov V, Agar T, Peretz J, Nafussi B et al. (2000). Effect of combined application of heat treatments and plastic packaging on keeping quality of 'Oroblanco' fruit (*Citrus grandis* L. *C. paradisi* Macf.) Postharvest Biol. and Tech. 20 (3): 287–294.

Sanchez-Ballesta BT, Alonso JM, Lafuente MT, Zacarias L, and Granell A (1996). Analysis of cold and conditioning-induced gene expression in 'Fortune' mandarin fruits. VIII Int. Citrus Congress, Sun City, South Africa (Abstract).

Schirra M, D'hallewin G, Carbas P, Agnioni A, and Garau VL (1998). Seasonal susceptibility to Tarocco oranges to chilling injury as affected by hot water and thiabendazole postharvest dip treatments. *J. Agric. Food Chem.* 46: 1177–1180.

Schirra M, D'hallewin G, Carbas P, Angioni A, Ben-Yehoshua S, and Lurie S (2000). Chilling injury and residue uptake in cold-stored 'Star Ruby' grapefruit following thiabendazole and imazalil dip treatments at 20 and 500° C. *Postharv. Biol. Technol.* 20: 91–98.

Tugwell BL and Gillespie K (1981). Australian experience with citrus fruits wrapped in high-density polyethylene film. *Proc. Int. Soc. Citriculture,* Int. Citrus Cong. Tokyo, Japan, vol. 2, 710–713.

Tuset JJ (1984). Postharvest citrus diseases in Spain: Problems and solutions. *Proc. Int. Soc. Citriculture,* Int. Citrus Cong. Sao Paulo, Brazil, vol. 2, pp. 508–510.

Tuset JJ and Marti MC (1988). *Sclerotinia sclerotiorum,* a new decay of citrus fruit in Spain. *Proc. Int. Soc. Citriculture,* VI Int. Citrus Cong. Spain, vol. 3, pp. 1451–1459.

Tuset JJ, Portilla MT, Hinarejos C, and Buj A (1992). *Aspergillus niger* and *Rhizopus oryzae,* Causing postharvest decay of citrus fruits. *Proc. Int. Soc. Citriculture,* VII Int. Citrus Cong. Acireale, Italy, vol. 3, pp. 1040–1043.

Whitney JD, Hartmond U, Kender WJ, Burns JK, and Salyani M (2000). Orange removal with trunk shakers and abscission chemicals. *Appl. Eng. Agric.* 16 (4): 367–371

Wills RBH, Ku VVV, Shohot D, and Kim GH (1999). Importance of low ethylene levels to delay senescence of non-climacteric fruit and vegetable. *Australian J. Exptl. Agric.* 39 (2): 221–224.

Wilson WC, Holm RE, and Clark RK (1977). Abscission chemicals—aid to citrus fruit removal. *Proc. Int. Soc. Citriculture,* Int. Citrus Cong. Orlando, Florida, vol. 2, pp. 404–406.

7

Diseases of *Citrus* and Their Management

Citrus is susceptible to a large number of diseases caused by different types of pathogens, namely fungi, bacteria, viruses, viroids, phytoplasmas, spiroplasmas, nematodes, etc. There are also a few diseases for which the causal organisms are still not properly known. Most of these diseases are systemic in nature while the infection caused by others is restricted to certain parts of the plant. Systemic diseases threatening to citrus cultivation are mostly graft transmissible but also transmitted by vectors. Most of the nonsystemic diseases, where pathogens are soil- or air-borne, affect the productivity and quality of production.

FUNGAL DISEASES

Fungal diseases are mostly nonsystemic in nature and infect the roots, foot/collar (crown), trunk (stem), foliage, and fruit. Common fungal diseases are: 1) Gummosis, 2) Anthracnose, 3) Scab, 4) Mal Secco, 5) Blight, 6) Root dry rot, 7) Powdery mildew. Several other diseases of minor importance are also found to infect citrus such as Greasy spot, Post-blossom fruit drop, Melanose, Black spot, and Brown spot. None of the fungal diseases is graft-transmissible.

Gummosis

Gummosis is a symptom indicated by oozing of gums from the infected part. This symptom may appear in roots and trunks due to infection by various fungi, e.g *Phytophthora, Diplodia,* and *Ceratocystis. Phytophthora* is the most prevalent among these pathogens. It also causes root rot, foot rot, collar-rot, and brown fruit rot.

Phytophthora Gummosis/Root Rot and Collar Rot/Brown Fruit Rot/ Blight

Phytophthora belongs to a lower group of fungi called Phycomycetes. The vegetative body, called the "*thallus*" is composed of richly branched coenocytic mycelia. Although it reproduces by both sexual and asexual methods, the latter method prevails and is responsible for its rapid spread. At the reproductive stage, it produces sporangia or conidia. It also produces chlamydospores to overwinter under adverse environmental conditions. Under favorable conditions, conidia germinate to produce biflagellate zoospores to form a new vegetative body. This genus has a large number of species. Some of these species are exclusively saprophytic or facultative parasites and survive in soil. A few are obligate parasites. Facultative saprophytic species under suitable conditions can infect a wide range of hosts, from trees to herbs. In some cases, root exudates induce chemotactic attraction to the zoospores. In all parasitic species, zoospores form the initial inocula.

Citrus is normally infected by five species of *Phytophthora,* namely: *P. citrophthora* (Smith et Smith) Leonian, *P. nicotianae* van Breda de Haan var. *parasitica* (Dastur) Waterhouse (=*P. parasitica* Dastur), var. *nicotianae* Water., *P. hybernalis* Carne, *P. citricola* Sawada, and *P. cactorum. P. nicotianae* var. *parasitica* is polyphagus. In addition to *Citrus,* it infects several herbaceous plants such as tomato, melon, and ornamentals. *P. citrophthora* var. *parasitica* is less polyphagous and may be parasitic on a very limited number of plants belonging to other families. *P. cactorum* is normally a competitive saprophyte. It may infect citrus when its population build-up is much more than that of other species. Parasitic *Phytophthora* may have several biotypes. *P. parasitica* is known to have three biotypes, depending on its virulence to different host species. All species cause gummosis, root rot and crown rot (foot rot). *P. hibernalis,* on the other hand, causes "brown rot" on fruits.

Inoculum Density and Disease Incidence

The population or inoculum density of all soil species is significant in making them parasitic. Generally the soil population of all *Phytophthora* species may increase in soil independent of the disease but depends greatly on soil type and other adaphic and biotic factors associated with the soil. If the soil is saturated, citrus trees become more susceptible to all the concerned *Phytophthora* species. Waterlogged soil predisposes a plant to *Phytophthora* infection. But there always remains a threshold level of population density above which infection occurs. Normally infection is caused by 2.66 propagules (colony forming unit) ml^{-1} soil (Anonymous, 1997). Lio et al. (1984). studied the inoculum density and disease incidence in citrus orchards and nurseries. They found different intensities of infection of root rot and blight (ranging from 15 to 80%) in different textural classes of soil from different localities (clay, loamy sand, sandy clay loam, and loamy sand containing different percents of organic matter, range 2.7–10.7, with more or less similar pH, range 6.0–7.9) with different inoculum density, ranging from 3–3.5 to 600 propagules g^{-1} of rhizosphere soil. Thus the threshold of inoculum density differs with soil type and soil organic matter content and also on the geographic location. So a nurseryman has to determine the level of inoculum content of the soil before raising a nursery or an orchard.

Symptoms

Gummosis: Appears at the trunk base, especially on the commercial scion part; the bark cracks and abundantly exudes gum. Subsequently, the bark color darkens and the internal tissues decay extending into the wood.

Foot rot or collar rot: Affected portion of the bark (canker) remains covered with soil leaving only a few centimeters of it above the soil surface. An outstanding hyperplasia of the canker border is present, both on the base of the trunk and on the principal roots.

In both diseases, the trees appear yellowish with pale green leaves with yellowish veins, smaller limbs, and short flushes; the branches partially or completely defoliated and wilting. When severe attacks occur, trees often die.

Blight/ brown rot of fruit: These are the symptoms on the foliage and fruit caused by aerial infection.

Varietal Susceptibility

Susceptibility of different *Citrus* species, varieties and cultivars to different *Phytophthora* species depends upon the soil condition and population density of the fungus in the soil. Among the rootstocks, Sour orange and Citranges are tolerant but Sweet orange, Lemons, Limes and related types are susceptible. Different scion varieties may modify the behavior of rootstocks, conferring higher or lower susceptibility to *Phytophthora*. Development of *Phytophthora* in citrus trees may also be modified by the occurrence of virus and viroid diseases and other stressing factors. Feichtenberger et al. (1992) evaluated the tolerance to *Phytophthora* species in different scion-rootstock combinations in Brazil. In their study, selection of Rangpur lime and *C. volkameriana* Palermo showed somewhat different reactions to *P. citrophthora* and *P. parasitica*; the former incites the production of larger lesions than those induced by the latter. *C. volkameriana* showed high resistance to *P. citrophthora* but not so high to *P. parasitica*. *C. karna* is more susceptible than *C. volkameriana*. The authors found Indian Rough lemon as the most resistant rootstock for both the fungi and *Poncirus trifoliata* very tolerant to them.

Natal sweet orange top conferred higher susceptibility to both fungi in cases of Red-ling-mung, Borneo red lime, Limeira Rangpur lime and Japanese Citron. Grapefruit scion confers susceptibility to *C. volkameriana* only against *P. citrophthora.* All scion varieties tested conferred higher susceptibility to *C. karna.* The hybrid Sunki mandarin *P. trifoliata* Swingle 311 was the most resistant one among the hybrids tested, both as a seedling and when grafted with Hamlin sweet orange scion. The authors tested 18 rootstocks for Hamlin sweet orange scion and observed a great variation in their performance against *P. citrophthora* and *P. parasitica,* so also with Piralima sweet orange and Cravo mandarin. Some appeared to be more susceptible while others appeared to be more tolerant or resistant. Variation in resistance was also noted using different rootstocks and Pera sweet orange scion.

These authors further observed *Citrus* tristeza virus (CTV) induced resistance to *P. citrophthora* and *P. parasitica* using preimmunized Pera sweet orange on Rangpur lime rootstock. The percentage of reduction was high for *P. parasitica.* Induction of resistance was greater with the virulent strains.

Phytophthora Management

The first logical step for management of *Phytophthora* in citrus nurseries is to preclude the introduction of this pathogen into the nursery. The important source of its entry is contaminated irrigation water and infected seeds. Seeds after extraction from the fruits are normally treated in water at 50° C for 10 minutes, dusted with a protective fungicide, and stored at 5° C. Growth media are made free from the fungus. If sand is used, it is fumigated with methyl bromide at 454 g per m. Containers in which seedlings are grown are freed of propagules by dip treatment in 10% formalin for 10 minutes. In case styrofoam seed trays are used, these are also fumigated under a plastic trap with methyl bromide. The seed trays are subsequently dipped in Styrodip, a copper containing plasticizer. Movement of infested soil and plants in the nurseries is avoided as far as practicable. Copper-containing footbaths at the entrances of various sections of nurseries and greenhouses are normally used to limit the access of *Phytophthora.*

Fungicide programs: Drenching of seedlings with furalaxyl at 0.62 g a.i. per 5 liters water per m is normally followed from the first day of planting of seeds on a weekly schedule. Infected seedlings are destroyed. Quintozene (PCNB) at 8 g a.i. per m can be incorporated into the surface of the seedbed-mix to control *Rhizoctonia* damping off. Benomyl may also be applied as a drench at 0.25 g a.i. per liter of water at the onset of emergence of seedlings to control *Rhizoctonia* damping off until the seedlings acquire resistance.

Phytophthora gummosis can be controlled by foliar sprays of efosite-AL at 3.2 g a.i. per liter of water every two months during the summer and fall seasons.

To control *Phytophthora* root/foot rot, propagule content in the orchard/container soil is to be regularly monitored. Samples are taken every two months. The number of samples should represent about 0.1% of the total number of trees. Sampling is done systematically and every four samples are mixed in a plastic bag. A 100 cc subsample of the soil is taken for analysis. The results are plotted in a map of the nursery/ orchard to delineate the infested areas. These areas are treated with fungicides when *Phytophthora* is detected in 30% of the samples. Treatment may be done with metalaxyl 0.02 g a.i. per liter of soil through irrigation water.

Before removing the trees from the nurseries, captafol at 0.01 g per liter of soil is drenched into the soil, if those plants are not be immediately planted.

Foot/root rot infected trees in small orchards or home yard gardens can be rejuvenated by inarching trifoliate orange rootstocks at foot region.

Brown Fruit Rot Management

Brown rot and green mold (*Penicillium digitatum*) are serious packinghouse diseases when picking is done in the rainy season, autumn or winter. Conventional method of control of these diseases are application of copper oxychloride, carbamates or Phalimides providing a fungicide barrier on the surface of the fruits to prevent them from infections by zoospores or sporangia. But these fungicides normally fail to provide effective control. Davino et al. (1992) studied the effect of Fosetyl-Al against these diseases. They observed that this fungicide when applied at 240 g a.i. per 100 liter as a foliar spray during October, November and December, assured proper control. It also improved the performance of trees and quality of fruits. They further found that this fungicide does not have any adverse effect on citrus mycorrhiza.

Fagoaga et al. (2001) successfully produced transgenic orange cv pineapple plants bearing a chimeric gene construct consisting of the cauliflower mosaic virus 35 s promoter and the coding region of the tomato pathogenesis related PR-5 protein. Challenging the regenerated transgenic lines by the pathogen, evidence is obtained for the in vivo activity of the PR protein against *P. citrophthora*.

Diplodia Gummosis

This gummosis is an aerial infection caused by *Diplodia natalensis* Pole-Evans. It belongs to Fungi Imperfecti that reproduces only by asexual methods. Its vegetative body is composed of haploid mycelia that are septate, profusely branched, and usually multinucleate. The septa are perforated and may permit streaming of the cytoplasm and movement of the nuclei from one cell to another. Reproduction is usually by conidial spores produced in a definite fruit body called the "pycnidium".

Symptoms
Gummosis and rotting begins at the stem end of the fruit but may also start at the stylarend or at injuries. Symptoms include discoloration around the bottom, abnormal pliability, and translucent watery rind similar to that of *Phomopsis* rot. The rot becomes dark, olive-green to black, especially on oranges. In a dry atmosphere, the fruit dries and the interior becomes a dark fibrous mass of tissues.

Management
This disease can be managed by the same method as applied for management of brown fruit rot.

Ceratocystis *Gummosis*
This gummosis is caused by *Ceratocystis fimbriata* Ellis et Halst.

Symptoms
Gummosis of the trunk: Gum runs abundantly to the base and wets the soil. On longitudinal incision, gum comes out of the puncture through little canals from the xylem. Branches turn yellowish and signs of nutrient deficiency appear in leaves. On transverse cut, concentric lines are found in the trajectories of the cambium of light chestnut color with a gummy look, including the rootstocks that apparently are not affected by this disease. Necrosis is found in secondary branches. Making successive cuts until reaching the wood, alternate spots of light chestnut color are found which get stronger the deeper they are. In the interior of stalks, a taint of salmon rose in a localized area of the wood the size of a pencil is frequently encrusted with gum. Under the fissured or healthy bark where the gum flow may be observed, the inner tissue is not affected and no necrosis occurs. Gum comes from the local punctures or bags through the little canals of the xylem. Sometimes bags do not appear. If cuts are made under the apparently healthy bark between the latter and the wood, a reddish gelatinous substance can be observed. In many cases, generally under the bark of the trunk, little points or protrusions emerge which usually coincide with a scaly or blistery surface.

Alcain and Marmelicz (1984) did extensive studies on this disease in lemon.

Gummosis in citrus does not necessarily imply an indication of a determined causal agent. It is a symptom of a reaction to the gradual deterioration of cells, more specifically the cell skin, manifested by the presence of gum in its tissue.

Most destructive gummosis is caused due to *Phytophthora*. Like *Ceratocystis* and *Diplodia*, several other fungi may also cause gummosis, such as *Sclerotinia* (an Ascomycetous fungus, producing ascospores in apothecia), *Diapothe, Dothiorella, Botryosphaera, Hendersonula,*

Cylindrocladium, Phoma, Corticium, Fusarium, Sphaeropsis, and others. Some viruses and pathogens of unknown origin also produce gummosis. These types of gummosis differ from that produced by *Phytophthora* in their quantity, extension, localization, and importance.

Anthracnose

Anthracnose on citrus fruit and foliage is caused by *Gloesporium limetticolum* Claus, which also belongs to Fungi Imperfecti and survives in soil. It is similar in vegetative structure to *Diplodia* but has a different type of fruit body called the "acervuli" which produces conidiospores.

Symptoms

Shedding of leaves and dieback of twigs are the main symptoms. On the dead twigs, the acervuli appear as black dots. This pathogen may also cause light green spots on leaves that later turn brown. The acervuli appear in concentric rings. The stem-end infection of immature fruits results in fruit drops. Necrotic effect may appear on leaves, young shoots, and tender fruits. Infected fruits develop a reddish-brown stain on the rind. Similar symptoms may also be produced by *Colletotrichum gleosporoides* belonging to the Fungi Imperfecti and reproduced through the production of acervuli.

Incidence of this disease depends on environmental conditions. Temperature between 34° C (maximum) and 21° C (minimum), with relative humidity above 90%, and rainfall favor the infection of this fungus. It normally affects new growth, preventing formation of fruit and producing lesions on fruits. A copper sulfate or other cupric fungicide applied during the rainy season at the first flowering or when the new growth appears normally controls this disease. Captafol, mancozeb or copper sulfate (Bordeaux mixture) gives satisfactory results. But repeated use of copper sulfate may increase the incidence of *Eutetranichus* mites.

Scab

Symptoms

Small semitranslucent dots that become sharply defined elevated pustules, flat or somewhat depressed at the center, appear on the leaves. In later stages, leaves become distorted, wrinkled and misshapen on the fruits. The lesions consist of corky projections that often break into scabs. Similar symptoms may also appear on fruits. The fungus can infect only under wet conditions.

This disease is caused by *Elsinoe fawcettii* and *E. australis* Bitanc et Jenkins. There are two biotypes of *E. fawcettii:* one is pathogenic to rough lemon, grape fruit and Murcot tangor but not to sour orange, sweet orange and Temple tangor. The other biotype is pathogenic to all these cultivars. *E. australis* so far is known to infect only sweet orange. Whiteside (1984) did comparative studies on the morphological, cultural and pathological characteristics of the two pathogens. According to him, there is no reliable host range to differentiate them. These two species are primarily differentiated on the basis of ascospore dimensions: 12–20 × 4–8 mm for *E. australis* and 10–12 × 5–6 mm for *E. fawcettii*. But one can hardly obtain the perfect stages of them for routine differentiation purposes. There are, however, several other features on which differentiation can be made. *E. fawcettii* has one feature in its anamorphic state that apparently is peculiar to this species: the production of a spindle-shaped colored conidia that have a *Cladosporium*-type development. But colored conidia are formed when the infected plants are placed in the outdoors for 1–2 weeks.

This disease can be avoided in nurseries by undertaking proper phytosanitary measures. It can be controlled by foliar spray of methyl thiophanate.

Mal Secco

This disease is caused by *Phoma tracheiphila* (Petri) Kanck et Ghic, of class Fungi Imperfecti. It usually reproduces forming conidia in a specialized fruit body called the "pycnidium". This fungus colonizes the vascular system and grows out of the xylem vessels only when the host collapses. There are two strains, a chromogenic (Pt_{55}) and a non-chromogenic (Pt_{42}) differing in their virulence. Pt_{55} is more virulent than Pt_{42}. These strains of *P. tracheiphila* are transmitted through infected fruits and seeds (Ippolito et al., 1987).

Symptoms and Mechanism of Development

The symptoms of 'Mal Secco' disease include chlorosis of leaf veins, yellowing and shedding of leaves, wood discoloration, dying of shoots and branches, and sudden wilting and collapse of the entire tree, mostly due to water stress. Thanassoulopoulos and Manos (1992) did mathematical calculations and statistics by the STATVIEW program on a Macintosh computer and developed a Model of the disease in *Citrus* orchards under normal conditions and evaluated the prognosis of epidemic development.

Mechanism of development. Magnano di SanLio et al. (1992) studied the relation between xylem colonization and symptom expression of Mal Secco.

Extensive vascular colonization is a prerequisite for symptom expression. Both pathogenicity and virulence of *P. tracheiphila* strains are dependent on their ability to colonize the xylem. The infection causes increase in electrolyte leakage, suggesting impairment of the cellular membrane permeability. This change precedes decrease in relative water content values. This alteration initially affects only the apical leaves showing symptoms; afterwards when the plant is about to collapse, it extends to basal leaves as well.

Extracellular pectic enzymes produced by *P. tracheiphila* are capable of inducing an increase in electrolyte leakage (Cacciola et al., 1990; Natoli et al., 1990), but their activity in the infected host tissue has yet to be demonstrated. Conversely, it is a widely held view that the pathogenic fungi produce toxins that alter membrane permeability, causing a water deficit. Parisi et al. (1992) demonstrated the phytotoxic activity of "Mallein", a low molecular weight metabolite of *P. tracheiphila*. According to them, specific toxic metabolite(s) are hydrosoluble, while the lipophilic fraction of the fungal culture fluid contains nonspecific toxins. The Mallein, a major component of the ethyl acetate extract of culture fluids of both chromogenic and nonchromogenic isolates of *P. tracheiphila*, could be responsible for necrotic symptoms induced by the ethyl extracts on tomato cuttings (100 mg ml^{-1}); conversely, the mallein may not be responsible for the phytotoxic activity of the crude culture filtrate of *P. tracheiphila* as it is present at a very low concentration. Cacciola et al. (1992) suggested possible involvement of two different polygalacturonase activities in the molecular mechanism of the pathogenesis of Mal Secco.

Deng et al. (2001) identified 15 genes related to the resistance of lemons to this pathogen by analyzing lemon suspension cultures treated with the fungal toxin using the PCR-select cDNA subtraction technique. The observation was confirmed by Northern blotting activated by toxin treatment. Such toxin activation of these was also detected in other genotypes.

Blight/Declinio

Citrus decline is often caused by several pathogens of unknown etiology. *Citrus* Blight is one such disease having great economic significance. This disease occurs in a large number of *Citrus*-growing countries, especially in the USA and Brazil where it has been termed as "Declinio". Other affected countries are South Africa, Australia, Caribbean islands, Venezuela, and Cuba.

Symptoms

The principal symptoms of Blight are similar to those associated with water stress. Earliest symptoms consist of rolling and hardening of leaves, and inter-veinal zinc deficiency in younger leaves. These symptoms gradually lead to wilting of affected foliages. Depending on the soil moisture and air temperature, wilted leaves begin to shed from the trees within several weeks or months of the first appearance of symptoms. Severe wilt and leaf loss is accompanied by dieback of affected branches. Blighted trees remain in a chronic decline condition and become unproductive. Vessel plugging is the most noticeable internal symptom in wood of the infected trees. Characteristically these plugs are strand-like or fibrous and composed of host-derived materials and may be of two different types, "lipid plug or resin" and "amorphous gum plug". There may also be bands of narrow vessels along the cambium. Gum plugs occur uniformly in old and new wood whereas resin plugs occur more commonly in older wood. Resin plugs predominate in rootstocks whereas gum plugs are predominant in scions (Nemec, 1985). Significant accumulation of zinc also takes place in the outer trunk wood, large limbs, and roots of the diseased trees. Albrigo and Young (1981) observed higher quantities of zinc accumulation in phloem of the trunk of the diseased trees.

Detection:

(1) Affected trees absorb much less water than healthy ones when water is injected in them. This feature is used to detect blight infection. Water can be injected both by gravitational methods or introducing it by injection using a syringe. To conduct the injection, a hole of about 3.2 mm diameter and 30 cm deep is drilled in the trunk, 20–30 cm above the bud union. Ten ml of water are injected within a period of 30 seconds. Respective water uptakes in healthy and blighted trees are 1.66 ml s^{-1} and 0.3 ml s^{-1}. Xylem plugging in blighted trees shows an inverse relationship with the water absorption.

(2) "*Wood Zinc Test*": Normally the zinc level is eightfold more in blighted trees than in healthy ones.

(3) *Dot-Immuno-Binding Assay (DIBA) and Western Blotting*: An antiserum to a 12 kd blight-related protein was found to be specific to trees showing blight symptoms (Derrick et al., 1990a, 1990b). This technique is also useful for detecting declinio/blight at the presymptomatic stage.

Fusarium and Blight

Fusarium solani Snyd. et Hans often occurs in the rhizosphere soil of citrus. In the field, *F. solani* causes fibrous and scaffold root rot on healthy and blight-infected trees but noticeably fewer symptoms occur on healthy-appearing trees (Nemec, 1984). Infection of roots can be initiated directly via enzymatic action through root cell walls (Nemec and Oswald, 1991) but entry may also occur through ruptures in walls of epidermal cells caused by movement of roots through sandy soil (Nemec et al., 1986). If the fungus does not expand infection into the cortex, it may reside in epidermal cells as chlamydospores (Nemec, 1978) and may represent a type of symptomless infection. Nemec et al. (1988) observed the production of naphthazarin toxin by *F. solani,* toxic to *Citrus.* This toxin was detected in *Citrus* xylem fluids from roots and branches in blight-diseased and healthy-appearing trees (Nemec et al., 1991). Nemec et al. (1988) reported that *F. solani* or its toxins increase vessel plugging in *Citrus.* Based on the observations made on toxins and their role in plugging in roots and outer barks, it has been postulated that during the symptomless period of a tree's life, periodic infections of roots by *F. solani* result in root loss and toxin released in the tree. Toxin translocation in the xylem causes resin and gum plugging to occur in roots and trunks. This continual build-up of plugging reaches a threshold after which the waterconducting tissues cannot meet the transpiration demand of the canopy (Nemec and Baker, 1992). Presence of the same toxin in the xylem fluid of *F. solani* infected and blight infected trees suggests a close association between these two diseases.

Van Rensberg et al. (2001) observed that *F. solani* produces naphthazarin toxins. A significantly higher concentration of these toxins is detected in blight-infrected trees. According to them, these toxins are involved in the induction of some of the symptoms associated with blight disease.

Botrytis Blight

Symptoms

The first symptom appears as twig dying that starts in the cut area and spreads downwards. Young shoots initially turn chlorotic, lose turgidity, and dry out, followed by dieback. Fruiting stage normally does not appear in the dead parts. Darkened and necrotic areas develop near the petiole on the main veins of some of the young leaves. Gummosis of the bark of young shoots sometimes occurs. Tip necrosis and irregular necrotic spots on mature leaves are observed in asymptomatic trees. Symptoms may vary in different grafting seasons. Symptoms normally develop in the protected nursery blocks (metal framed tunnel covered by plastic sheets and shading nets) and were first reported by Polizzi and Azzaro (1992).

This disease is caused by *Botrytis cineria* Pers. Ex. Fr., the conidial stage of an ascomycetous fungus called *Sclerotinia fuckeliana* (de Bary) Fekl. Conidiophores of the fungus produce aggregated conidia of variable size and shape; microconidia may also occur with conidia. Approximately 106 conidia are required to induce infection.

Root Dry Rot

Symptoms

The most conspicuous symptom found above ground is the fatal collapse of the tree; leaves suddenly wilt and dry out within a few days, remaining attached to the tree. The course of the disease is normally chronic; trees show a lingering decline progressing for years, showing slight wilt under dry condition, with defoliation and poor vegetation flush. There may be twig dieback, chlorosis and yellowing of main leaf veins, and leaf fall. Infected trees often show delayed blooming and an unusually large crop of fruits. These symptoms are common to those caused by agents affecting roots or girdling the trunk. Specific symptoms of this disease are visible at the base of the trunk. Patches or large areas of the bark at ground level appear moist and dark; affected bark gradually dries but eventually cracks but still adhere to the wood. Wood under the dead bark remains firm, showing dark or dark-brown discoloration. In terminal cases, wood of the crown becomes spongy starting from the taproot and roots decay leaving only a few adventitious roots. Unlike *Phytophthora* root rot, in dry rot there is no gum oozing (Polizzi et al., 1992).

Primary causes of dry root rot are not definitely known. *Fusarium solani* is normally considered the primary pathogen (Menge, 1989). But manifestation of the disease is always associated with other biotic and/ or abiotic stress factors. Overwatering, poor drainage, poor soil, etc. contribute to its outbreak.

Powdery Mildew

Symptoms

White powdery growth appears on young twigs and especially on the upper surfaces of leaves. Infected leaves turn yellow and gradually dry. In case of acute infection, infected and dried leaves fall, causing dieback of the twigs. The fungus may also appear on the skin of young fruits and cause premature fruit drops.

The pathogen of this disease is an ascomycetous fungus called *Oidium tingitaninum* Carter. It is an obligate parasite. Its mycelium is hyaline and septate. It is common on sweet orange, sour orange, mandarin and Rangpur lime. Infection normally takes place by air-borne spores called conidia. These are unicellular, hyaline, barrel-shape wth one slightly tapered pole, borne in chains on short hyaline conidiophores. Rate of production of the conidia is very high, but their germinability is very low (approximately 50%). Germination of the conidia depends upon environmental conditions in particular temperature and relative humidity. The rate of germination is high within a temperature regime between 15–20° C. Conidia normally fail to germinate at or below 2° C and at or above 35° C. The optimum relative humidity for the germination is between 80–100%. The conidia usually do not germinate below 60% relative humidity. On germination, hyphae enter directly into the host cells and not through any stomata. But the conidia do not remain viable for a long period. They usually remain viable for 15 days. However, the fungus may survive in blackish mycelial form on the undersurface of old leaves, forming distinct zones or black pustules. When these zones or pustules from 6-month-old leaves are put to favorable conditions, new growths appear. The survival of this fungus in mycelial form has been recorded in *C. reshni, C. sinensis, C. aurantifolia,* and *C. limonia. Poncirus trifoliata* has been found to be an alternative host to this fungus.

There is variability in susceptibility of different citrus species to this fungus. The spread of this disease correlates significantly mild to minimum temperature and relative humidity (Dutta Roy, 1994; Dutta Roy et al., 1991).

SOME OTHER DISEASES OF FOLIAGE AND FRUIT

Greasy Spot

Symptoms: Yellowish brown to black lesions appear on the undersurface of leaves. Minute black lesions may also appear on fruit rind. Infection of an ascomycetous fungus called *Mycosphaerella citri* causes this disease. It over-winters in infected decomposed leaves on the ground. Under favorable temperature and humidity, the fungus produces ascospores. These spores are carried by wind and cause infection. The spores germinate on the undersurface of leaves and penetrate through stomata. It may take several months for the appearance of symptoms.

Postbloom Fruit Drop

Symptoms: Orange brown lesions appear on the petals of infected flowers; infected fruitlets often drop but the calyces and floral disks often remain attached to the stems. The disease is caused by *Colletotrichum acutatum* (class Fungi Imperfecti). It overwinters on resistant structures, infected leaves, and floral parts. The fungal mycelia produce conidia that are carried by rain splash or wind. These conidia germinate under favorable conditions to cause infection. Symptoms may appear within 4–5 days after infection.

Melanose

Symptoms: Raised, brick-red to black lesions appear on twigs, leaves, and fruits. Pliable, leathery stem-end rot may also occur usually beginning at the bottom or cut end of the fruits in orchards or storages; the rot varies in appearance according to the maturity of the fruit, temperature, and moisture.

The disease is caused by *Phomopsis citri* Fawcet (class Fungi Imperfecti). It produces green or dark thick-walled pycnidia bearing ovate or fusiform air-borne spores. These spores under favorable conditions cause infection. The fungus overwinters in infected dead twigs. The perfect stage of this fungus is *Diaporthe citri* belonging to Ascomycetes. This stage produces ascospores in perithecia that remain embedded in dead twigs and may disseminate by wind or rain splash. These spores normally do not infect. On germination, they produce the Imperfect form of the fungus that in turn produces infective conidia.

Kuhara (1999) developed a simulation model "MELAN" to provide a system analysis of the ecology of the causal fungus, the infection process, and disease development, as well as the protective effect of applied chemicals.

Black Spot

Symptoms: This disease produces black lesions of various size on fruits and leaves of susceptible species. An ascomycetous fungus called *Guignardia citricarpa* causes this disease. At its reproductive stage, the fungus produces sunken, spherical and black perithecia that contain only asci; paraphyses are usually absent; asci are clavate, eight-spored. The ascospores are hyaline, elliptic or fusiform. The fungus overwinters in infected tissues. The spores are airborne and germinate on leaf and fruit surface and cause infection. Symptoms normally develop several months after infection, particularly at the time of maturity of fruits or even after harvest.

Alternaria Brown Spot

Symptoms: Dark lesions are produced on stems, leaves, and fruits on infection of susceptible cultivars. The symptoms on fruits which may appear both in orchards and storage are normally characterized by: (1) decay extending into the center of the fruit accompanied by breakdown or collapse; (2) semipliable end; and (3) dry, firm, black rot at the navel end. In case of lemons, the breakdown of the central axis or core extends upward from the bottom or end. The rind splits, and juice or the disintegrating tissues pop out under slight pressure. In the case of oranges, a firm, dry, slow form of decay occurs in the stylar end, resulting in destruction of the cells of the pulp sacs.

This disease is usually caused by a species of Fungi Imperfecti, namely *Alternaria citri* Ell. et Pierce, emend. Bliss et Fawcett. Several other species may also be associated with this disease. The fungus under certain conditions produces a toxin that may rapidly kill the leaves and twigs causing dieback of the entire shoot.

The fungus reproduces by forming airborne conidia in the lesions, which are mostly obclavate or oval, beaked or nonbeaked, muricate with mostly two to four transverse septa and one to four longitudinal, light to olive-brown in color. The fungus overwinters in infected tissues and the conidia germinate and cause infection. Symptoms may appear within 36 hours if the conditions are favorable.

BACTERIAL DISEASES

Bacterial diseases infecting citrus are very limited in number. But there are two internationally threatening diseases called *Greening* and *Citrus Canker.* Greening is systemic and graft-transmissible whereas Canker is nonsystemic and the symptoms remain restricted to the infected plant parts. Recently, a xylem-limited bacterium was found to be responsible for *Citrus Variegated Chlorosis* (Rossetti et al., 1990).

Greening

Greening implies the greening of fruits as the principal malady. This disease has been devastating in countries where citrus was domesticated from primary or secondary forests of Asia, Africa, and Arabian countries. It is indigenous in these countries and was known by different local names. In China it was known as "Huang lung bin"(HLB)/"yellow shoot"; in Indonesia,"vein phloem degeneration"; in Taiwan,"likubin"; in Philippines, it was " leaf mottling"; and in India,"dieback" or "decline". This disease was first scientifically discovered in South Africa and designated "Greening". Subsequently, similar symptoms were observed and recorded as "Greening" in several other countries, in particular Pakistan, West African countries south of the Sahara, Nepal, India, Re-Union Island and others. Considering the normal transmission of this disease by contaminated planting materials, it is conjectured that its spread occurred in various countries from certain focus points. It is believed that Greening was introduced in the Philippines, Thailand, and Malaysia from China. As India has a long history of this symptom since 1920, it is further believed that China and India might be the major foci for the distribution of Greening in Asia and the Arabian peninsula.

Global Scenario

Greening is a serious limiting factor for citrus production in Asia and Africa (Aubert, 1992). According to Aubert's estimate, dozens of million of trees have been lost in China and Taiwan since 1950 and the economic life expectancy of newly planted orchards considerably reduced until the 1970s when coordinated efforts were deployed to eradicate the disease and launch a new plantation policy in vector-free areas with certified disease-free planting materials. According to Ke and Xua (1990) and Chen (1991), half a million trees were removed to implement the eradication campaign and to bring back the production and productivity in China. In

Indonesia, citrus economy was severely affected due to Greening since 1960 with an estimated loss of three million mature trees and far more number of young nursery plants up to 1970 (Tartawidjadja, 1980). This disease almost wiped out the best commercial Keprok mandarin in Java. It is believed that Greening gradually spread throughout this archipelago from a single focus identified in the early 1950s near Jakarta Pasar Mingu. A vast eradication campaign was initiated in Indonesia to clean the planting materials by shoot-tip grafting and promoting the development of registered disease-free nurseries (Becu and Whittle, 1988; Winarno and Supriyanto, 1991) along with new planting programs in vector-free areas. In the Philippines, it has been estimated that between 1960 and 1970 more than five million trees were affected by "Leaf mottle" disease reducing the area of cultivation by 60 percent. To date this disease remains as an endemic one in most of the southern islands.

In Africa, severe crop losses by a disease showing greening symptoms were recorded as early as 1932. It is thought that greening is responsible for the loss of millions of trees in South Africa and the loss gradually spread to East Africa up to Ethiopia in the North, the upland areas of Cameroon in the West, and to Madagascar in the East (Aubert et al., 1988). This disease hinders citrus production in the best horticultural uplands in several other Afican countries such as Malwai, Burundi, Rwanda and the local markets remain undersupplied in various citrus fruits except for a few acid limes and rough lemons.

Symptoms

Greening pathogen produces diverse symptoms depending on the fruit type, extent of infection, season of observation, and nutrient status of the orchards. In general, considerable stunting, leaf and fruit drop, and twig dieback occur. There may be out-of-season growth flushes and blossoming. Infection and appearance of symptoms may be sectorial, occurring only in a portion of the tree or only a branch or a twig. When the entire tree is affected, it becomes stunted, sparsely foliated, with chlorotic green fruit, poor crop, and poorly developed root system with relatively few fibrous roots.

Typical Greening is invariably accompanied by a great diversity of foliar chlorosis of which a type of mottle resembling zinc deficiency often predominates (Plate II: Figure VII.1). There may also be chloroses resembling those resulting from deficiencies of iron, manganese, calcium, sulfur and boron. There is also reduction in leaf size. Initially young leaves from new growth appear to be normal, showing only a pale green color. They gradually tend to assume an upright position, become leathery, develop a dull olive-green color and the veins become prominent by turning yellowish.

Fruits normally remain underdeveloped, usually lopsided, small, poorly colored, with curved columella, color unequally distributed, sides exposed to direct sunlight normally develop full color whereas the remaining portion appears dull, olive-green, with only a touch of orange. When pressure is exerted on such fruits, a grayish white, waxy deposit may appear on the surface of the rind. Infected green fruits often develop high shoulders at the stem end and thick puffy rinds; vascular bundles stand out prominently in the albedos, usually yellowish-brown in color.

Both foliar and fruit symptoms are prominent in mandarins, oranges, tangers, tangelos, grapefruits and pomelos. In these plants the appearance of symptoms after infection may take four to six months. Limes and lemons are less sensitive and seldom show prominent symptoms, remaining symptomless reservoirs. Hung et al. (2001) identified an alternative host of the greening pathogen called Chinese box orange (*Severinia buxifolia*). The pathogen could be transmitted to this plant both by grafting and vectors.

Organism

The Greening organism was initially labeled as virus because of its characteristic symptoms, and vector and graft transmissibility. Later, it was described as a "Mycoplasma". In contrast to mycoplasmas and spiroplasmas, however the Greening organism, though polymorphic and flexible, contains an outer layer or coat. Comparative study of the greening agent and several gram-negative plant pathogenic bacteria showed the greening agent is to be a "Bacteria like Organism" (BLO). The sensitivity of the greening agent to penicillin also indicated its bacterial nature. But the absence of peptidoglycane (PG), found in the bacterial cell wall and conferring definite shape and rigidity to it, apparently separates the greening agent from the bacteria. Garnier et al. (1984) established that the greening organism is a "Gram-negative Bacterium". They studied the morphology of its structure and found that the envelope surrounding the organism comprised three zones: (i) a dark inner zone, (ii) a dark outer zone, and (iii) an intermediate electron-transparent zone. The thickness of the three zones is approximately 250 Å. Each of the two dark zones was resolvable into a triple layered unit membrane 90–100 Å thick. The inner membrane appeared as a cytoplasmic membrane and the outer one as a cell wall. On transferring the BLO to periwinkle and allowing it to multiply, they obtained morphologically and structurally similar BLO in phloem tissues of the infected periwinkle. Applying the cytochemical treatments developed to visualize the PG layer of *Escherichia coli*, they successfully demonstrate the presence of PG in the envelope of the BLO and confirmed its true bacterial nature. Sui (1988) claimed success in purifying the Greening organism showing identical structural and serological features to prokaryotic Gram-negative bacteria. He found the shape of this organism to be as oval or elongated with occasional constrictions and varying size (30–600 nm x 500–1400), nm, the coat thickness 30–70 nm versus only 7–10 nm in Mycoplasmas. Garnier et al. (1987, 1991) obtained monoclonal antibodies (mAbs) against two Asian BLO strains (Poona, India and Fuji, China) and one African strain (Nelspruit, South Africa). These mAbs are highly strain specific and so far seven different BLO serogroups have been identified (Gao et al., 1993). Garnier and Bove (1993) isolated a purified Greening organism using monoclonal antibody (mAb) and confirmed the observation of Sui (1988). Villechanoux et al. (1992) cloned and sequenced a 2.6 kbp DNA fragment (In-2.6) of an Indian (Poona) BLO genome. This fragment corresponds to the rather well conserved *rplKjL-rpoBC* bacterial operon and in particular, codes for four ribosomal proteins (L1, L10, L11, and L12). When, for taxonomic purposes, the sequence of the BLO operon was compared with the sequences from other bacteria, the HLB-BLO was unambiguously identified as a member of the eubacteria. But a comparison of the protein sequences deduced from the genes with their counterparts in other bacterial species failed to reveal any specific relationship between the BLO and any previously described bacterial species. Fragment In-2.6 was able to hybridize under high stringency conditions with all the Asian strains but not the African strain when Southern hybridization was conducted. At lower stringencies, the hybridization between In-2.6 of the Asian strain with the African strain was possible, revealing DNA polymorphism. Sequence comparisons between the PCR-amplified, cloned, and sequenced 16S ribosomal DNAs of the Asian strain (Poona) and the African strain (Nelspruit) revealed that the two BLOs are members of the class Proteobacteria (Jagouex et al., 1994). Considering molecular and phytogenetic observations, Murray and Schleifer (1994) proposed the designation "*Candidatus*" as an interim taxonomic status to record the sequence-based potential of new taxa at the genus and species level. This possibility was used in the case of HLB-BLOs by naming the African HLB-BLO "*Candidatus* Libobacter africanum" and the Asian HLB-BLO "*Candidatus* Libobacter asiaticum"(Jagouex et al, 1994; Garnier and Bove, 1997).

Transmission and Spread

Greening is a graft-transmissible disease and contaminated bud/rootstocks mostly serve as primary source of inoculum. Normally 50% of such grafts show typical foliar symptoms. The pattern of its field spread indicates that the major mode of the transmission of greening is through vectors. A *Citrus* psyllid (*Trioza erytreae* Del Guercio) was a suspected vector in South Africa as early as 1965 and McClean and Oberholzer (1965) actually confirmed its vectorial role that year. Individually both male and female insects can transmit the pathogen. On the other hand, Matsumoto et al. (1961) while studying the "Likubin" disease in Taiwan had already established another *Citrus* psyllid (*Diaphorina citri* Kuway) as the vector of that disease. This pathogen is a circulative one in *D. citri.* A vector becomes infective after an incubation period of 21 days from acquisition and can retain the pathogen throughout its life. The acquisition period is short. These authors further demonstrated that if the pathogen is acquired by the psyllid in the 4th and 5th instar stage, the emerging adult is automatically infective.

In contrast, *T. erytreae* requires no incubation period and becomes infective after access to an infected plant for 24 hours; nymphs of this species are not infective and the pathogen persists in an adult for 2–3 weeks. It appears from the contrasting mode of transmission that the greening pathogens may be of two types, Asian or Oriental, transmitted by *D. citri* or the African greening, transmitted by *T. erytreae.* Asian greening is heat tolerant (good symptoms at 30 –35° C) and the African greening heat sensitive (no symptom above 27 –32° C).

Xua et al. (1988) conducted further studies on the transmission of Citrus Huang lung bin by *D. citri* and confirmed transmission of the pathogen by 4th and 5th instar nymphs and not by 1st to 3rd instars. They found the minimum latent period of the pathogen in the insect as one to two days and the maximum latent period as 25 days after acquisition feeding. The pathogen could be located in the cells of salivary glands, in the filtration chamber of the foregut and cells of mid- and hind-gut. As they are able to retain the pathogen, once acquired, these insects can transmit it to a large number of plants. But they do not move a long distance as they are not efficient fliers and mostly are incapable of infecting an orchard 2 km away from the focus of infection. The population build-up of these insects is seasonal and the abundance of eggs and nymphs often correlates with the emergence of new flushes (Weerwut et al., 1987). The most interesting aspect of vector-greening-citrus is that the presence of citrus and the vector does not always mean the presence of greening. Aubert (1985) found the Asian psyllid in Brazil and Okinawa but not greening. This author (1986a, 1986b) also found *T. erytreae* in Gabon and Cape Province but no greening.

There are climatic and altitudinal differences in the distribution of these two psyllid species. *T. erytreae* does not establish freely in arid or semi-arid climate with low rainfall and high temperature, but occurs in subtropical climate at an altitude of 1,000–2,000 m (Aubert, 1987). *D. citri,* on the other hand, is more tolerant to climatic extremes. It can be found from sea level up to an elevation of 1,500 m (Chao et al., 1979; Aubert et al., 1985; Aubert, 1984; Lama et al., 1987). Mukhopadhyay et al. (1992)

studied greening and *D. citri* in Darjeeling hills where greening occurs mostly due to infective planting materials and the occurrence of the psyllid remains restricted to certain pockets.

These insects can be found in hosts other than citrus but only in plants belonging to Rutaceae. Egg laying and nymphal development of *T. erytreae* is restricted to 15 and 13 species respectively, whereas *D. citri* can breed on 21 species (Table 7.1).

Table 7.1: Hosts of *Trioza erytreae* Del Guercio and *Diaphorina citri* Kuway

Nature of hosting	*D. citri*	*T. erytreae*
Preferred host	*Murraya paniculata*	*Clausena anisata*
	Citrus aurantifolia	*Vepris lanceolata*
		Citrus limon
		Citrus medica
		Citrus aurantifolia
Common host	*Citrus limon*	*Citrus sinensis*
	Citrus sinensis	*Citrus nobilis*
	Citrus medica	*Citrus reticulata*
	Citrus nobilis	*Citrus deliciosa*
	Citrus reticulata	*Citrus paradisi*
	Citrus deliciosa	*Citrus grandis*
	Microcitrus australiana	*Murraya paniculata*
	Citrus paradisi	*Fragrara capense*

In addition to these preferred and common hosts of these psyllids, there are a number of occasional hosts.

Occasional hosts of *D. citri*: *Citrus hystrix, Citrus grandis, Triphasia trifoliata, Fortunella* sp., *Poncirus trifoliata, Murraya koenigii, Todalia asiatica, Vepris lanceolata,* etc.

Occasional hosts *of T. erytreae*: *Dodalia asiatica, Fortunella* sp., *Poncirus trifoliata, Calodendron capense, Microcitrus australiana,* etc.

Detection

Although greening, as a disease has been known since nineteenth century, discovery of the morphological and physical structure of it is only a recent event. Thus its detection methods extend from simple biological and histochemical to sophisticated serological and molecular methods. Some of the biological and histochemical methods still have relevance to detection at the field level, where sophisticated equipment is not available for precise detection at the molecular level.

Biological method: Use of indicator plants is the most widely used technique to detect greening at the field level. Sweet orange is mostly used as the indicator plant. The suspected plant is used as the scion and sweet orange as the rootstock to conduct graft-transmission and characteristic symptom development. Vector transmission is also done for the same purpose. But these processes are very time consuming. Sometimes it takes several months for symptom development.

Histochemical methods: Schwarz (1968) developed a technique for rapid diagnosis of greening pathogen. In this method, chromatograms prepared from leaf and bark extracts by thin layer chromatography are examined under "Hanovia" ultraviolet lamp at a wavelength of 360–366 nm (>95% UV light) where fluorescence can be observed in the chromatograms prepared with the extracts from the diseased plants. Chromatograms show a purple spot at an Rf of 0.5 to 1.0 just next to the spotting points.

This method can be routinely used for initial diagnosis of greening in sweet orange and mandarins but not in pomelos and acid limes. There are also a number of other histological methods to detect greening. These methods include direct detection of fluorescence after staining with aniline blue and observation of sections after staining with safranin. Wu (1987) used the direct fluorescence detection method to diagnose citrus yellow shoot disease in China. In this method, freehand transverse section of petioles mounted in water are directly observed under a reflecting fluorescence microscope when bright yellow fluorescence is found in the phloem tissues of the infected petioles. Abnormal fluorescence in phloem tissues in freehand sections of midribs of infected leaves can also be observed by staining the sections with 0.1% aniline blue and observing them under a fluorescent microscope (Wu and Faan, 1988). The cheapest and most handy method is the staining of freehand transverse sections of the infected petiole with 0.1% safranin for two minutes and observing the red patches in the phloem tissues under an ordinary light microscope.

Electron microscopic method: This method was in use before the development of sensitive serological methods. Ultrathin sections from midribs of infected plants are fixed with 4% glutaraldehyde, postfixed with 2% osmium tetroxide, dehydrated in an acetone series, and embedded in Spurr's resin. The sections are then stained with 4% uranyl acetate and 2% lead citrate and observed under an electron microscope where the greening pathogen can be seen in the sieve tubes (Prommintara, 1988).

Serological methods: These methods are conducted using monoclonal Antibodies (mAbs) for the pathogen. Garnier et al. (1987) first produced the mAbs of the Indian greening (Poona strain). Subsequently mAbs for the Chinese strain were produced (Garnier et al., 1991). These antibodies were used for serodiagnosis of different strains of greening by the double-sandwich DAS-ELISA technique. These scientists produced two hybridomas (2D12 and 10A6) for Indian greening (Poona strain) and assayed the respective antibodies by differential avidin-biotin ELISA and Immuno-Fluorescence (IF) with sections of periwinkle midribs. On the contrary, they found four hybridomas (2H9, 6GI, 11H6, 12E12) for Chinese greening. The 2D12 and 10A6 monoclonal antibodies specific for Poona greening could detect all Indian isolates. These antibodies recognized two different epitopes on 22,000 Da antigenic protein of Poona greening, suggesting the occurrence of different antigenic structures on the surface of different greening organisms. The chemical nature of these structures may differ. Some epitopes may be protein while others may be carbohydrate. Thus greening pathogens showing morphological similarities may be of different serotypes. These sero-types may occur in close proximity within a country (Garnier and Bove, 1993). Interestingly, mAb 10A6 can recognize strains from different countries. Using this mAb, an antigenic protein was purified by immuno-affinity chromatography and named *P42*. It is a surface protein evenly distributed on both the round and filamentous forms of the greening pathogen. This protein can produce the mAb by in vitro immunization of spleen cells. In addition to 10A6, it also produced two new mAbs having the ability to recognize more strains. One of them, mAb 1A5, recognized all the Asian strains the authors tested except those from China and Africa. Thus several strain-specific and broad-spectra mAbs are available for general and highly precise detection of different sero-types of greening pathogen.

Nucleic acid hybridization: This technique was developed for rapid and precise detection of greening. Using the cDNA, it is now possible to detect greening in trees and individual psyllid.

Garnier and Bove (1993) purified total phloem DNA extracted from periwinkle plants infected with Poona greening. This DNA they digested with restriction endonuclease Hind III and the resultant DNA fragments were cloned in Replicative Form (RF) using the phage M1 3 mp 18 as vector and three selected recombinant phages p3, p19, and p10 containing the Poona pathogen DNA inserts, by differential hybridization against DNA extracted from healthy and greening infected periwinkle midribs. The inserts of p3, p19, and p10 were 2–6 kbp, 1.0 kbp and 0.6 kbp respectively in length and were designated as In 2.6, In 1.0, and In 0.6. When the cloned DNA fragments were used as probes in Southern or dot hybridization experiments done at high stringency (60c, 0.1SSC), the smallest insert, In 0.6 hybridized only with DNA of periwinkle or citrus plants infected with homologous Pooa strain. In 1.0 gave positive hybridization signals with all Asian greening strains tested except the Taiwan strain while In 2.6 hybridized with all Asian strains tested. Under these conditions none of the probes hybridized with DNA extracted from healthy plants or with the DNA extracted from the African strain. Bove et al. (1993) and Varma et al. (1993) used In 2.6 as a probe for detection of the greening pathogen in orchard trees in India. Electron microscopic detection of the greening pathogen was also performed on the same samples. These observations confirm that In 2.6 is more sensitive than EM for detecting greening pathogen.

Greening Management

Greening disease is caused by a graft-transmissible phloem-restricted bacterium and its field spread occurs by a psyllid carrying the bacterium in a persistent manner. Thus the management of this involves the following operations: (i) preclusion of the primary inoculum by planting greening-free grafts/seedlings, preferably in vector-free areas, (ii) prevention of entry of vectors into orchards, (iii) regular monitoring of infective vectors in orchards and (iv) control or management of vectors in case of their chance arrival. A knowledge of epidemiological features of the greening pathogen is necessary. Aubert (1992) outlined some of these features (Table 7.2). (i) Greening-free planting materials can be produced by the shoot-tip grafting technique. (ii), (iii) and (iv): see Chapter 5: *Citrus* Production Technology.

Table 7.2: Some epidemiological features of Asian and African greening

Features	Endemic Asian type	Endemic African type
Suitable climate	Warm and cool day and equatorial climate	Humid/cool climate
Preferential host plants for the vector	Cultivated and ornamental plants belonging to Rutaceae	Common shrubs or trees of the natural habitat
Opportunities of vector-free areas	Frequent, especially near forest area	Infrequent, heavy infestation near forest area
Dispersal activity vector	Flying-landing activity on short distances, 0.5–1.0 km to 2.3 km	Flying-landing activity on medium distances, 3–4 km or more
Disease pattern	Aggressive, mostly disseminated; near cultivated or ornamental Rutaceous plants	Widespread above a given latitude and constantly present to a greater or lesser extent
Disease spread	Mostly through the transport of the pathogen and the vector within contaminated plants	Equally by the vector and the operator
Eradication program	Successful with tight prevention measures	Difficult to implement

Cultural Management

Some aspects of cultural management have been described in Chapter 5: *Citrus* Production Technology. Cultural management of greening primarily involves preclusion and management of vectors to keep them below threshold level in orchards. Such management practices differ in Asian and African greening where the nature of the vector species differs.

African greening: Neither the pathogen nor the vector thrive under warm conditions of lowland areas but are common in cool humid climates above 500 m elevation. The vector *T. erytreae* develops primarily on indigenous Rutaceous plants, especially *Vepris lanceolata* (Lam.) G. Don, *Zanthoxylum capense* (Thumb.) Harv., and *Clausena anisata* (Willd.) Hook. F. ex. Benth, the latter, being a common shrub in forest and bush steppes of East and West Africa. Therefore forest situations in Africa support recurrent *T. erytreae* colonies that are sources of permanent re-infestation and the insect is able to fly distances of 10–12 km or more. Thus management starts with selection of sites for orchards. In Africa, sites should be located in lowland areas with warm climates below 200–300 m elevation and far away from forest land.

In raising orchards in highlands in cool humid climate, nurseries are to be enclosed in greenhouses under insect-proof conditions, and orchards raised with proper windbreaks and under a rigorous protective umbrella with chemical sprays of triazophos, aldicarb or monocrotophos, especially during the spring flush corresponding to vector outbreaks (Catling, 1969). Vector population in orchards can also be brought down below the threshold level by adopting biocontrol methods as demonstrated by Aubert and Quilici (1984).

Two eulophid ectoparasites (superfamily Chalcidoidea, order Hymenoptera), *Tetrasticus dryi* Waterson and *T. radiatus* Waterson were introduced, bred and released in Reunion Island. Populations of both the vectors of African and Asian greening were drastically reduced 24 months after an original rate of release of 30–40 eulophid ectoparasites per square km of the citrus area. These ectoparasites can strongly limit the vector population. Such populations can be effectively maintained in orchards.

Asian greening: Both the pathogen and its vector *D. citri* thrive in hot climates with temperatures well above 30° C as well as in highland cooler climates, although to a lesser extent. Thus it is difficult to locate vector-free areas. But *D. citri* develops only on a few domesticated plants, *Citrus* and *Murraya,* and can fly only a short distance (2–3 km). *D. citri* is not found in natural forest habitat but is largely disseminated in densely populated rural areas on citrus and *M. paniculata* backyard trees. The countless home citrus plants are normally unsprayed and constitute most dangerous reservoirs of both the vector and the pathogen. As the forests are devoid of the vector, these situations can be used for resetting new disease-free orchards of Foundation Blocks (Ke and Xua, 1990).

Selection of site is also an important item for the management. Nurseries are to be raised in isolation (>2 km away from any orchard), preferably inside an insect-proof greenhouse. Orchards are also to be planned in safe isolated areas far away from villages and sprayed during spring flush that corresponds to *D. citri* outbreaks. Each orchard is to be kept under regular vigil for incidental increase of the vector and the foci are to be identified and eradicated with immediate effect by spraying dimethoate or monocrophos. Special attention is to be given during March to May, before the monsoon when the population increases to a great extent. If the coordinated eradication program is undertaken from the beginning and implemented along with adequate vector control, orchard sanitation can be achieved and the disease incidence precluded.

Citrus Canker

Citrus bacterial Canker is another disease of common citrus species. It is now threatening citrus production in particular limes and lemons throughout the world. The incidence of this disease is also found in other citrus cultivars. It has been recorded as endemic in many Asian citrus-growing countries and introduced from these endemic areas to other citrus-growing countries in particular South America, Africa, Australia, and Oceania. Normally this disease does not occur in arid regions; but the threat persists through introduction of infected planting materials from other areas. An Eradication Campaign of great magnitude was initiated in Florida, USA in 1910 and continued up to 1933 when the state became canker-free. No canker-infected material was found in Florida up to 1984 when a new form of canker appeared. Another eradication campaign was begun and continued up to 1991 to free the area of that form of canker (Stall and Civerolo, 1991). A similar eradication program and restrictive regulatory measures were undertaken in Brazil in 1957, but a new outbreak occurred there in 1979. Similar situations have occurred in various countries (Rosetti et al., 1982; Civerolo, 1981, 1984; Stall and Civerolo, 1991; Goto, 1992).

Like "greening", incidence of citrus canker may also have been influenced by domestication of citrus plants from primary/secondary forest habitats or transfer of germplasms from one habitat to another. It too is an ancient disease known by different names in different countries and the respective causal agents were also recorded by different names. The oldest record of citrus canker dates back to the middle of the nineteenth century when lesions were detected in citrus samples maintained in the Herbaria of the Royal Botanic Gardens, Kew, UK (on *Citrus medica*, a wild forest plant from India collected during 1827–1831 and *Citrus aurantifolia*, a domesticated plant collected from Indonesia during 1942–1944). Canker has also been reported from a wild forest plant *Fortunella hindsii*. Subsequently, citrus canker has been reported from different countries, infecting different citrus and the related hosts; causal pathogens have been recorded as *Phytomonas citri*, *Pseudomonas citri*, *Xanthomonas citri*, etc. Hasse (1915) identified the pathogen as *Xanthomonas campestris* pv. *citri*. Currently, it has been renamed as *Xanthomonas axonopodis* pv. *citri*. Till 1981, three different forms of this pathogen were recognized (Goto et al., 1980; Stall et al., 1980). Namekata and Oliveira (1972) reported a new form of canker from Mexican lime and Rodriquez et al. (1985) reported another new form from the same plant. Thus it is apparent that the pathogen is gradually mutating to evolve new forms under different host-pathogen-environment situations. With the advancement of citriculture and with the development of new cultivars/hybrids/mutants of different citrus species, the scope of evolution of the canker pathogen is probably also increasing, extending more threats to the health of citrus crops.

Symptoms

Citrus bacterial canker disease produces different symptoms in different hosts. Its characteristic symptom is occurrence of conspicuous lesions on leaves, twigs, and fruits. Severe infection also results in defoliation, dieback and premature fruit drop. Mitra (1995) observed that typical lesions occur on all hosts but size differs from 1.2 mm to 5.07 mm depending on the host species/cultivars; it is smallest in acid lime and largest in grapefruit; lesions in some cases may have a 'halo' and some may produce black exudates on the abaxial surface of the infected leaf.

Forms of the Pathogen

Various workers (Goto et al., 1980; Stall and Civerolo, 1991) have described the five different forms of Bacterial *Citrus* Canker Disease (CBCD). The most common form of canker is "Asiatic canker", "True canker", "Canker A" or "Cankerosis A" (CBCD-A). It was first recognized in Florida. It affects most *Citrus* species and hybrids as well as other Rutaceous and non-Rutaceous plants. Grapefruit, trifoliate orange, sour orange, sweet orange, Mexican lime, lemon, mandarin, tangerine, and tangelo are severely affected by this form. Mexican lime is highly susceptible to the Asiatic type; it also produces mild symptoms on a few sweet oranges and Tahiti lemon.

"Cankerosis B", "Canker B" (CBCD-B) was first reported from Argentina. In contrast to CBDB-A, it is difficult to culture. It primarily affects lemons, but Mexican lime, sour orange, Rangpur lime, sweet lime, citron and occasionally sweet orange may also be affected. "Mexican lime Cankerosis" or CBCD-C and "Bacteriosis Canker" or CBCD-D are so far known to affect only Mexican lime, in contrast to A and B which can infect all *Citrus* species. CBCD-C was first reported from Brazil in 1972. It was found to differ from A and B in serological properties and resistance to bacteriophages CP 1 and CP 2 A and B are susceptible to these phages whereas C is resistant to them. It also shows characteristic symptoms. CBCD-C forms begin as small water-soaked areas on young and succulent leaves and twigs; on leaves, lesions enlarge and become light tan to brown raised pustules surrounded by distinct chlorotic halos; older lesions become concave with a crater-like appearance on the upper surface and convex on the underside. Older necrotic lesions eventually collapse, becoming corky and often cracked. The diameter of a lesion usually ranges from 2–7 mm (Sanches-Anguino and Felix-Castro, 1984).

CBCD-D was first reported from Mexico in 1981 (Rodriguez et al., 1985). It produces symptoms similar to A and is serologically related to A and B. It is very weakly pathogenic and differs in host range. In addition to Mexican lime, it infects Persian lime, lemon, grapefruit, and other species in nurseries. CBCD-E was first reported from Florida in 1987 (Schoulties et al., 1987). It was initially called as "Bacterial Spot of *Citrus*". This canker differs from A, B, C and D in lesion formation producing corky and raised lesions; lesions caused by E are flat to sunken below the leaf surface. There is, however, controversy over whether this isolate should be treated as a separate form of canker; some authors prefer to record it as D (Gabriel et al., 1989).

Characteristics of the Pathogen: X. axonopodis pv. *citri* is a rod-shaped, Gram-negative, aerobic motile bacterium with a single polar flagellum. Cell size of different forms does not significantly differ and its normal dimension is 1.5–2.0 × 0.5–0.75 nm. It has smooth surface, round ends, entire margin with convex elevation, and appears translucent. Its growth is moderate, lusturous with botryose consistency, pale yellow color and distinct odor. Cells are encapsulated and do not produce endospores.

The bacterium utilizes glucose with oxidative reactions, ferments several other sugars, related compounds, and several derivatives. Some of these are arabinose, mannose, lactose, trehalose, xylose, sucrose, maltose, starch, dextrin, malonate, citrate, succinate, and malate. It fails to utilize fructose, galactose, L-arabinose, rhamnose, raffinose, glycerol, methyl glucoside, salicin, inositol, dulcitol, mannitol, sorbitol, inulin, gluconate, oxalate, acetate and tartarate. Utilization of mannitol varies among the isolates. The culture produces ammonia and hydrogen sulfide and can liquefy gelatin. It does not respond to methyl red test but responds to peptonization and potato soft rot test, the degree of which varies with the type of canker.

The culture in nutrient broth, pH 7.2, shows a distinct growth pattern with an initial stationary phase of 0 h to 2 h followed by a lag phase of 0 h to 4 h, and exponential phase of 4 h to 12 h, negative growth acceleration phase of 12 h to 40 h, maximum stationary phase of 12 h to 28 h, acceleration death phase of 28 h to 48 h. All forms show similar growth patterns but their response to biochemical tests and pathogenicity differs (Rodriguez et al., 1985; Goto et al., 1980). Form C fails to use maltose, lactose, mannitol and malonate; Form B can use mannitol while Form A can use all of them.

Iwamoto and Oku (2000) identified the essentiality of *hrpX* gene of the pathogen for its pathogenicity. They cloned and analyzed the homolog, designated *hrpXt*. Its *ORF* has 1,431 *bp* in nucleotides having a coding capacity 476 amino acid residues with a molecular mass of 5.24 kDa. According to the authors the structure of the *hrpX* gene is highly conserved in the genus *Xanthomonas*. Almeida et al. (2000) observed the presence of proteins bound to *Ap 11*, a virulence factor of *Xanthomonas campestris* cv. *citri* (*X. axonopodis* pv. *citri*) required for the formation of canker symptoms on *Citrus* in the fraction 25–50% ammonium sulfate column. Western analysis revealed that *Ap 11* binds specifically to *25 Kd* and *110 Kd* proteins. It was also found that *Ap 11* targets exist specifically in citrus plants.

Detection

Symptoms, although providing the first cue to the presence of canker, can pose a problem since many other diseases, pests or injuries often produce similar symptoms. Furthermore no specific biochemical and physiological tests exist for reliable detection and identification of the pathogen. There is also no specific medium for this bacterium. Thus the common and dependable method is determination of its host reaction or pathogenicity test. Goto (1992) described a very simple technique called the "Leaf infiltration technique". Samples to be tested are washed in water and the washings first centrifuged at low speed (2,000 rpm for 10 min) to remove coarse particles and plant residues, and then at high speed (9,000 rpm for 20 min.) to precipitate bacterial cells. The sediments are suspended in a small amount of sterile water and infiltrated with a syringe into the mesophyll of the leaf blade of a young plant grown in artificial medium under aseptic conditions. Mitra (1995) found that lesions appeared within 10 days when the inoculum was injected to the host cell. The population of the causal bacterium in the sample could be calculated from the number of lesions developed on the basis of an average water infiltration rate of 7 $\mu l\ cm^{-2}$. The minimal detection efficiency by this technique is about 10^2 bacterial cells g^{-1} soil. The number of lesions that develop is directly proportional to the concentration of the bacterium between 10^2 and 10^5.

But this method is time consuming and it has a built-in risk in infiltrating the inoculum in the test plant.

Serodiagnostic tests based on the occurrence of form-specific, heat-stable antigens were developed to achieve a more sensitive, risk-free, and rapid detection method (Massina, 1980). Applying this method, prior isolation of the pathogen may be precluded. Subsequently, sensitive and efficient ELIZA methods were developed for routine detection of distinct forms. Baoquing et al. (1991) found that the sensitivity and specificity of detection were higher by double antibody ELISA than indirect ELISA. They could accurately detect canker from over 100 samples from 11 countries within 40 h. They further found that employing SPA ELISA detection could be done in 6 h.

A DNA fragment of *X. axonopodis* pv. *citri* has also been identified and sequenced, making it possible to produce a primer to develop a sensitive detection tool by PCR (Pruvost, 1992; Hartung et al., 1993). This fragment is absent in genome of all the citrus saprophytic bacteria and most plant pathogenic *Pseudomonas.* More recently an immuno-capture nested PCR Protocol (IC-N-PCK) associated with colorimetric detection of amplified DNA was optimized. This technique is now routinely used in laboratories for diagnostic purposes (Hartung et al., 1996). With this technique, the threshold level of pathogen detection has been improved by a factor of 10,000 compared to the conventional ELISA technique. The latest improvement of the technique consists of a "single tube nested PCR procedure" (Pruvost et al., 1997).

Management

Integrated management of citrus canker involves comprehensive understanding of the survival and spread of the bacterium, relative susceptibility of the different citrus cultivars coupled with adequate quarantine regulations and sanitary measures. The bacterium may survive in parasitic, epiphytic as well as in saprophytic form. It persists as a parasitic form for a limited period of time in diseased plant tissues; in epiphytic form on host and nonhost plants, and as saprophytic forms on straw mulch and soil (Goto, 1992). The pathogenic potential of these forms and also the lesions differ with different cultivars, age of the plant, and various conditions.

Soil is not an important source of the bacterium, as it does not survive in natural soil. But it can survive at least two to three months in leaves and twigs that have fallen to the ground and remain dry. Another source of natural infection is straw mulches. When the bacterium is splashed from lesions onto straw, it can survive there for several months under cool and dry conditions. In uncontrolled orchards, there occurs a clear cycling of the bacterium through different seasons, overwintered in bold lesions on leaves, twigs and barks, and reinfecting new flushes in spring and autumn. There may also be latent infection in some cultivars. Goto (1992) reported the role of weeds in overwintering of the bacterium. According to him, this bacterium can survive for a long period on *Zoysia japonica* and *Calistegia japonica*. These weeds are immune to the bacterium. It normally arrives through rain splashes and is sustained on the rhizomes or root systems of these grasses during winter or early spring when the pathogen can hardly be isolated from holdover cankers on *Citrus* trees.

The approach to management differs in countries where canker is endemic. In the endemic area, the first campaign is "Eradication" or "cut" and "burn" of diseased trees and trees adjacent to them. Then replanting of disease-free materials, along with strict regulatory measures, prohibiting entry of any infected sample or fruit and finally adopting appropriate sanitary measures. This strict Eradication and cultivation program was pursued in Florida and Brazil with much success. Canker was first introduced to Texas in 1910 through the introduction of infected trifoliate orange seedlings shipped from Japan and had become widespread in Florida by 1914. An Eradication campaign was then conducted at a cost of 56 million US dollars, destroying 257,745 orchard trees and 3,093,110 nursery plants and the United States became canker-free in 1947 and remained so until 1984 when another such campaign had to be undertaken after the appearance of a separate form of canker. *Citrus* canker has also been eradicated from Australia, New Zealand, and South Africa. In Xingguo county of Jiangau province of China, over 6 million superior disease-free (including canker) seedlings were raised in 27 years from 1950 and a set of Plant Quarantine Rules for producing an Area of Citrus Nursery Stock was approved as the National Standard in 1985 and extended to ten more Provinces (Baoquing et al., 1991). Luyi and Chengdong (1991) developed a set of rapid eradication techniques of *Citrus* canker without destroying infected trees or seedlings and found them highly effective. The operations involved (i) forcing leaves to drop and killing the bacterium; (ii) pruning, training, and sanitation of orchards; (iii) overturning down the surface soil; (iv) smearing or painting a chemical agent on the trunk; (v) spraying bactericide to protect shoots; and (vi) quarantining. Based on these procedures, they successfully cured 650,000 infected seedlings and 10,000 infected trees within a short period of time without destroying either.

The above methods are also applicable for management of *Citrus* canker in orchards. In this case, management is to start from the selection of disease-free seeds, as the bacterium may be seed-borne in orange, citron, and lemon (Singh et al., 1995). While raising the nursery rootstocks, attention should be given to keep the nursery seedlings free from the bacterium. This bacterium may produce small lesions at the basal portion of the leaf blade at the junction of the leaf petiole or on stems.

In view of the fact that the bacterium is mostly wind-borne, the orchards are to be fenced by a proper windbreak. The protective effect of a windbreak covers a distance six times greater than its height. With windbreaks, wind inside *Citrus* groves should preferably be maintained at 6 m s^{-1}. Lines of windbreaks are made perpendicular to the trade winds and encompass orchard blocks of 2 to 5 ha. Wind break trees are to be planted at least 8 m away from the first line of *Citrus*. Common windbreak trees are *Casurina* sp. In humid tropics, common windbreak trees are *Leucaena glauca* and *Erythrina fusca*.

A leaf miner (*Phyllocnistis citrella*), is a common carrier of the canker bacterium and the damage caused by this insect facilitates canker infection; hence effective control practices against leaf miner are to be pursued. Periodic spraying of nicotine sulfate or isoxanthion may control infestation of leaf miner (Goto, 1992; Gottwald et al., 1997).

According to Goto (1992), an outbreak of *Citrus* canker can be forecast on the basis of (i) number of overwintered lesions on angular shoots, (ii) temperature, (iii) frequency and amount of precipitation, (iv) number of windy days in October and November, and also (v) temperature in winter. Analyses of these data make it possible to predict the canker outbreak one or two months in advance and to schedule chemical control methods to preclude it. Application of Bordeaux mixture before the first flush in late March followed by applications of copper compounds with calcium carbonate or antibiotics every week is recommended. Application of Bordeaux mixture in June and August or whenever strong winds and heavy rains are predicted, reducing the concentration of copper by one half and increasing the concentration of lime to 1.5 times, can also be done to protect the leaves and fruits from chemical damage. Actual intervals for spraying copper compounds are to be determined according to the cumulative amount of precipitation.

Pruvost et al. (1997) conducted spatiotemporal analyses of CBCD in simulated nurseries. Spatial autocorrelation analyses were done with the LCOR 2 software (Gottwald et al., 1992). The observed aggregated disease patterns with a lack of directional spread, indicating splash dispersal of inoculum by irrigation water up to a distance less than one meter from the source. From their detailed epidemiological studies, they recommended an improved production scheme which is as follows: (a) mother plants for clean bud-wood should be grown in an insect-proof greenhouse; (b) budded plants grown in plastic bags should be grown in tunnels and watered by drip irrigation, to avoid free water on leaves, which promotes infection; (c) bud-wood stock should be screened for asymptomatic infection by IC-N-PCR. They also conducted trial on the durability of this clean planting stock strategy and observed that the durability may be broken by medium or long-distance transportation of the bacterium due to extreme climatic events such as tornadoes/hurricanes, etc.

Fu et al. (2001) observed that the spread of the canker (*X. axonopodis* pv. *citri*) correlates with temperature. Spraying with different fungicides showed that a 500-fold and 400-fold solution of 77% copper hydroxide and 60% chlorothanolin solution respectively could efficiently control the disease.

Variegated Chlorosis

Citrus variegated chlorosis (CVC) was first recorded in Brazil in 1987 on sweet orange leaves and fruits. The characteristic symptom of this disease is the appearance of chlorotic variegation on leaves and fruits. Rossetti (1990) did an electron microscopic study of the infected samples of leaves and fruits and found the agent to be a xylem-limited bacterium. He first suspected it to be a strain of *Xylella fastidiosa*. Chagas et al. (1992) confirmed his observation. This bacterium was isolated and artificially cultured in synthetic growth media (Chang et al., 1993). It was rod shaped, 1.3 μm in length and 0.2–0.4 μm in diameter with a rippled wall. An antiserum against an isolate of the bacterium gave strong positive reactions in double-antibody-sandwich (DAS) enzyme linked immunosorbent assay (ELISA) with other cultures from CVC-infected samples (Chang et al., 1993a; Garnier et al., 1993). When the cultured CVC bacterium was inoculated in sweet orange seedlings, it became detectable by DAS-ELISA within 3 months after inoculation and systemic within 4 months after inoculation when it could be reisolated; typical symptoms appeared 6 months after inoculation (Chang et al., 1993a).

Xylella fastidiosa is known to cause several other diseases in particular Pierce's disease of grapevine (Wells et al., 1987). Observations made by various workers indicate that the isolate from sweet orange is a strain of the same bacterium. Like isolates from other hosts, citrus isolate remains restricted to xylem tissues; all the isolates have similar morphology both in plants and in vitro, and can be grown in the same types of growth media. All the isolates are serologically related and have similar protein patterns in polyacrylamide gel electrophoresis. CVC-specific antisera show positive reactions with isolates from other hosts. Serological relationship between the CVC-isolate and isolates from other hosts suggest it belonging to Group I strains (Garnier and Bove, 1997).

In Brazil, *X. fastidiosa* infection is also found in coffee, causing leaf scorch or scalding (Bertha et al., 1996). The coffee isolate is similar to CVC-isolate. Li et al. (2001) mechanically inoculated a triple cloned strain of *X. fastidiosa* isolated from coffee and *Citrus* in coffee plants. They detected the pathogen 3 months after the inoculation both by ELIZA and PCR. Both inocula could be reisolated and showed the same characteristics as the original one by microscopy, ELISA, and PCR, and produced typical CVC symptoms.

Transmission

The sharpshooter (*Homalodisca coagulata*) is an efficient vector of *X. fastidiosa* for peach and grape. This insect also feeds on citrus. In Brazil, the sharpshooters *Acrogonia terminalis, Dilobopterus cotalimai,* and *Onchometopia fascialis* occur regularly on *Citrus* and have been shown to be vectors of the CVC-bacterium (Lopes et al., 1996; Roberto et al., 1996). He et al. (2000) observed transmission of *X. fastidiosa*-inoculated plants sharing the same pot and confirmed its transmission by root grafting.

Goldschmidt (2000) observed the susceptibility of tangerines to CVC. They also evaluated the resistance in 15 tangerines or mandarins and hybrids. Tangerines and their hybrids Wilking, Fortune, Sunki, Ellendale, Orlando tangelo, Nunes clementine, Nova, Sun shu shakat, Suencut and Batanges showed leaf symptoms with positive reaction to ELISA and PCR. In tangerines Cravo and Oneco, no symptom was observed suggesting their tolerance to CVC. In tangerine Dancy and mandarins Okitsu Satsuma and Ponkan, no symptom developed nor could the pathogen be detected suggesting their resistance to CVC.

Biological Detection

Monteiro et al. (2001) identified Madagascar periwinkle (*Catharanthus roseus*) cv. *peppermint* Cooler as an experimental host plant for the *Citrus* strain of *X. fastidiosa* strain 9a5c by mechanically inoculating the plant with the pathogen. Typical symptoms of leaf deformation and stunting appeared within 2 months after inoculation. The presence of the pathogen was tested by the PCR. Systemic infection caused dysfunction of plant growth within 4 months. Presence of the pathogen in the xylem was verified by immunofluorescence. Genes coding for protein with homologies to plant sterol-c-methyltransferase, a transkitolase-like protein subunit III of photosystem I, and a desiccation-protectant protein were differentially expressed in symptomatic *C. roseus* plants as a response related to pathogenesis.

Pathogen: Genomes and Genomic Diversity

Citrus variegated chlorosis is a xylem-limited fastidious bacterium. Simpson et al. (2000) completely sequenced the genome of this bacterium. They found the genome to contain 2,679,305 base pairs (bp) 52.7% GC-rich in a circular chromosome and two plasmids of 51,158 bp and 1,285 bp. It was possible to assign putative functions to 47% of the 2,904 putative coding regions. The mechanisms associated with the pathogenicity and virulence were found to involve toxins, antibiotics and ion-sequencing systems as well as bacterium-bacterium and bacterium-host interactions mediated by a range of proteins. At least 83 genes are bacteriophage derived and include virulence-associated genes from other bacteria, providing direct evidence of phage-mediated horizontal gene transfer.

Qin et al. (2001) isolated, cloned and sequenced a 5823 bp cryptic plasmid from a strain of *X. fastidiosa* and constructed a shuttle vector to transform the bacterium.

Silva et al. (2001) also completely sequenced the genome of the bacterium. Analysis of the genomic sequences revealed a 12 kb DNA fragment containing an operon closely related to the gum-operon of *Xanthomonas campestris* excluding three genes, but synthesizing a different EPS. It was suspected that this novel EPS (exopoly saccharide) consists of polymerized tetrasaccharide repeating units assembled by sequential addition of glucose-1-phosphate, glucose, mannose, and gluconic acid on a polyphenol phosphate carrier.

Ciapna and Lemos (2000) characterized the DNA of the isolate *9a5c*, amplifying it by PCR using primers corresponding to repetitive sequences present in the bacterial genome (*REP*, *ERIC*, and *Box*) were used. The isolate was identified by DNA amplification with the specific primer (*CVC-1/272-3 int/RST 31/RST33* ± *CVC1/CVC2*) generating 500 bp, 733 bp, and 270 bp fragments respectively. The haplophytes obtained by rep-PCR showed bands with molecular weights ranging from ~ 400 bp to 4,000 bp.

Qin et al. (2001) evaluated the genetic diversity of *X. fastidiosa* isolated from diseased citrus and coffee using repetitive extragenic palindomic PCR and interbacterial repetitive intergenic consensus PCR assays. They failed to distinguish any difference between the isolates from those plants. When comparisons were made to reference strains isolated from other hosts, *four groups could be identified. These were: (i) Citrus and coffee, (ii) grapevines and almond, (iii) mulberry, elm, plumb, and (iv) oak.* Independent results from random amplified polymorphic DNA (RAPD) PCR assays were consistent with these groupings. Two of the primers tested in a RAPD-PCR were able to distinguish the coffee and the *Citrus* strains. Sequence comparisons of a PCR product amplified from all strains confirmed the presence of *c f o I* polymorphism that can be used to distinguish the *Citrus* strains from all others.

Mehta et al. (2001) assessed the genetic diversity of the *X. fastidiosa* isolated from citrus in Brazil by r-PCR-RFLP of the 16s r-RNA and 16s-23s intergenic spacer and r-PCR fingerprinting. They obtained two major groups within the *Citrus* cluster and relationships to the geographic origin of the strains.

Spiro-plasmal Diseases (Citrus Stubborn Disease)

Stubborn is an obstinate dwarfing disease of citrus. This disease was first reported from Palestine as early as 1931 and then from California, USA. Top working with budwood from thrifty trees failed to improve the diseased trees, so the propagators called this disease *Stubborn*. Later, incidences of similar symptoms were recorded in Florida, Texas, Arizona (USA), Morocco, Mediterranean basin, Brazil, Argentina, Peru, Algeria, France (Corsica), Turkey, Syria, Israel, Jordan, Saudi Arabia and Iran (Bove et al., 1988). But it was known by different names in different countries. Some of these names are: acorn, blue albedo, pink nose, stylar-end greening, puny leaf, and little leaf.

Symptoms

Stubborn infection produces a wide variety of symptoms. Most or all parts of the tree may be affected; normal branches often occur in the diseased tree and milder symptom may appear in the shaded portion. Fruits become small, malformed, lopsided with a curved columella, or acorn-shaped, with a thick peel on the stem end and thin peel on the stylar end, irregular or inverse coloration (stylar end greening), premature mummification, white waxy appearance of the rind where pressure is applied; insipid or sour taste; and early fruit dropping. Growth flushes become out of phase; blooms several times a year. Growth becomes abnormal with shortened intervals between leaves, bunchy, upright, with an excessive number of shoots, multiple buds, small cupped leaves, variety of chlorotic or mottled leaf patterns resembling various nutritional disorders, green or premature yellow banding of veins of some leaves, twig dieback, etc. The most characteristic symptoms of the disease, however, are the abnormal branching and foliage and excessive seed abortion. Blue albedo may not be a specific symptom of this disease. The principal fruit symptoms can be induced experimentally in many varieties under controlled greenhouse conditions. Symptoms can be produced in different varieties of sweet orange, tangelo, mandarin, grapefruit, citrumelo, sour orange, rough lemon, Eureka lemon, Mexican lime, Palestine sweet lime, Rangpur lime, trifoliate orange, etc.

Inoculation of the pathogen is most effective when young stem tips and very young leaves are used as inoculum.

Pathogen

Considering the graft transmissibility of the stubborn disease, it was initially regarded as a virus or a transmissible stunting factor. Later, it was identified as mycoplasma-like organism culturable in artificial media and the Koch's postulate for its confirmation as a pathogen of the stubborn disease could be demonstrated.

In culture, the morphology of the pathogen closely resembles that of the mycoplasmas infecting animals. It shows "fried egg colony" morphology on agar; it is resistant to penicillin but sensitive to tetracycline; its RNA polymerase is resistant to rifampicin; it does not revert to walled forms even after 10 passages in penicillin-free media; it can be isolated and cultivated in the total absence of any antibiotics. But the striking properties of the ultrastructure of the stubborn pathogen appear in liquid culture. It appears essentially as a beaded and branched filament. Filaments are helical in form and motile, showing two types of motility—a rapid rotary motion and a slow undulation and bending of filaments. Its motility is visible in its colony morphology in that the umbonate colonies are diffuse, often with satellite colonies developing from the foci adjacent to the initial site of colony development. Nonmotile variants are also known. These colonies exhibit a typical umbonate appearance with sharp margins and without satellite colonies. The stubborn pathogen lacks flagellum. It is rounded by a unit-membrane of cytoplasm without a definite cell wall. There is a layer of surface projections adhering to the membrane called "nap". This nap has some resemblance to the outer surface of the triple layer outer envelope (periplast) that surrounds the protoplasmic cylinder of many Spirochaetes. Filaments also occur on which certain regions are free of both the outer nap and an innermost layer exposing the remnants of the protoplasmic cylinder.

The stubborn organism is Gram-positive (mycoplasmas are Gram-negative). A classic, tailed bacteriophage attached to and develops within this organism (not found in mycoplasmas). It is not serologically related to any known mycoplasma. These unique features led to its identification as *Spiroplasma citri gen. nov.* sp. nov. (Cole et al., 1973; Saglio et al., 1973; Markham and Townsend, 1974), the name drived from its helical (spiral) structure. Although all isolates from different origins of citrus plants show a similar pattern, slight variations as occur between them, indicating the occurrence of mild and severe strains of the stubborn pathogen (Mouches et al., 1980).

Besides chromosomal (genomic) DNA, *Spiroplasma citri* has extrachromosomal DNA: plasmids and/or viral DNA (Bove et al., 1988). The plasmid content varies from one strain to the other, certain strains having more than one plasmid. Four plasmids of *S. citri* have been characterized. These plasmids are useful for characterization of different strain of *S. citri* and also differentiating them from other *Spiroplasmas.* Under certain conditions, plasmid DNA may become integrated with chromosomal DNA.

Spiroplasma citri may also contain viruses. Three viruses—SpV1, SpV2 and SpV3—have been isolated from *S. citri.* SpV1 is a nonlytic, rod-shaped virus (ca 230 × 15 nm) with singlestranded circular DNA of 8.0–8.4 kilobase (kb). SpV2 is a polyhedron (55 × 50 nm) with a long noncontractile tail (ca 80 × 7 nm); its DNA is probably double-stranded. It may be specific to *S. citri.* SpV3 is a polyhedron (ca. 40 × 36 nm) with a short tail (ca. 15 × 7 nm); DNA is double-stranded and linear (Bove, 1988). The exact relationship between the DNAs of *S. citri,* its plasmids and viruses, and their occasional integration is not yet properly understood.

Detection

Stubborn pathogen is mostly detected by indexing on sensitive plants at suitable temperature (32 –35° C). Isolation and culturing of the pathogen in artificial media are also pursued to record the typical characters of the *Spiroplasma citri.* Serological methods are also employed to detect the pathogen. The most convenient method is ELISA (Saillard et al., 1980). The technique gives consistent results with *S. citri* infected periwinkles even when the plants are grown at temperature lower than 32° C, the optimum growth temperature of *S. citri.* This technique is able to detect 10 ng of *S. citri* proteins. Leafhoppers can also be screened for *S. citri* by this technique.

Transmission and Control

Stubborn pathogen is transmitted in fields by leafhoppers. Three species of leafhoppers are responsible for its transmission. These are *Circulifer* (*Neoaliturus*) *tenellus, Scaphytopius nitrides,* and *S. acutus delongi.*

The pathogen can also be cultured from infected leafhoppers. Oldfield et al. (1976) collected leafhoppers from several locations in California and assayed for the presence of *S. citri* in them by feeding them on indicator plants (Madagascar periwinkle) and by culturing *S. citri* from their bodies. They used the serological deformation test to determine the apparent relationship and identities of selected cultures of Spiroplasmas from field-collected *C. tenellus.*

Among the eight leafhopper taxa collected and fed on periwinkle, these authors found only *C. tenellus* to be frequently infective. The rate of transmission from *Scaphytopius nitridus,* a recognized vector of *S. citri,* was only 0.7%, which indicates its inefficiency for field-transmission of the disease in Californian conditions. *S. acutus delongi* failed to transmit. *S. citri* could be cultured from the field samples of *C. tenellus* (25%) and *S. nitridus* (4%).

S. citri could be cultured in *S. acutus delongi* (100%), *Ollariana strictus* (88%), *Acertagallia curvata* (75%), *C. tenellus* (50%), and *S. nitridus* (50%). When the indicator plants were inoculated after three weeks or more of acquisition feeding with the cultures, the leafhoppers showed a higher rate of transmission (*C. tenellus,* 77%; *S. acutus delongi,* 64%; *S. nitridus,* 29%). *A. curvata, Graminella sonora, Euscelidus variegatus, Microstele fascifrons,* and *O. strictus* could acquire and retain *S. citri* by feeding on infected periwinkle but failed to transmit, though *S. citri* could be cultured from field-collected *O. strictus.*

Oldfield and colleagues (1984) further found that transmission occurred late in the nymphal stage or in early adulthood following acquisition by young nymphs. Either sex can transmit the organism and there was no transovarian transmission. Minimum accession period was found to be 2 hours and the minimum transmission access period was found to be 6 hours. The minimum latent period in adults was 17–18 days. This period often lasted for several months during the winter. According to Bove and Garnier (1997), in the Old World two major leafhoppers, *Circulifer haematoceps* and *Circulifer tenellus* transmit the stubborn disease while in the New World, only *C. tenellus* transmits this disease.

Control

Treatment of budwood for 90–120 minutes at 51° C (water saturated hot air) can inactivate the stubborn pathogen. Inactivation in budwood can also be done at 50° C for 3–4 hours, although occasionally it survived the treatment. Pre-conditioned materials (kept first for 28 weeks at a temperature between 30° and 42° C) showed a considerably higher percentage of graft survival. Heat inactivation followed by shoot tip grafting ensured almost complete removal of the pathogen.

PHYTOPASMAL DISEASES (WITCHES'-BROOM)

Witches'-broom disease is commonly found in lime and occasionally in other species. Witches'-broom disease of lime (WBDL) was first seen in the Sultanate of Oman in the 1980s. It was seen in the United Arab Emirates (UAE) in 1989. It was next reported from South East Iran in 1997. In India, it was reported as a new disease of lime in 1999. Bove and Garnier (2000) studied this disease as found in all these countries except India and described the pathogen as *Candidatus Phytoplasma aurantifolia,* which is closely related to phytoplasmas of alfalfa, sesame and sunhemp. They produced monoclonal antibodies and PCR primers for the 16s rDNA amplifications obtained from the Omani Phytoplasma which could detect the incidence of the same pathogen in UAE and Iran. Leafhopper vector *Hishimonas phycitis,* multiplying on lime trees assumed to be the vector of this pathogen.

VIRAL AND VIROID DISEASES

Viruses are infectious nucleoprotein molecules. Their infectivity mostly depends on their structured nucleic acids, Deoxyribonucleic Acid (DNA) or Ribonucleic Acid (RNA). Viroids, on the other hand, are only infectious-free nucleic acids having no protein coat. These nucleic acids are similar to those operating the vital functions of all living cells. Thus viruses are capable of remaining integrated to any cellular system occurring in nature. Considering the ancient nature of *Citrus* plants, it may be presumed that several viruses and viroids had peacefully co-existed in the cellular systems of these plants for so long that they became components of primary and secondary forests. As soon as domestication and propagation by grafting of *Citrus* began, segregation of these infective molecules took place and they became aggressive pathogens. As the plants were brought out of their natural habitats, cultivated in different agroclimatic conditions and changes made in their genomics of their hosts to make them more productive, new forms of viruses also emerged in several cases. Because of this dynamics, *Citrus* virologists, at the beginning of their journey with viral diseases, found utmost difficulties in isolating the viruses in pure forms until sophisticated technologies were developed in the 1980s.

Viral diseases of *Citrus* began to be recognized in the 1930s. Diseases for which no microscopic pathogens could be detected but which were graft transmissible and producing characteristic symptoms, used to be recorded as viral diseases named according to the hosts and the characteristic symptoms they produced. Some of the diseases recorded from 1939 to the early 1970s are: Xyloporosis, Infectious variegation, Psorosis, Exocortis, Cachexia, Satsuma dwarf, Vein enation, Tristeza complex/Lime dieback/Hassaku dwarf/Quick decline, Seedling yellows, *Citrus* mosaic, *Citrus* leaf curl, Yellow vein, Impietratura, Tatter leaf, Variola, *Citrus* ringspot, Natsu dai dai dwarf, Cristacortis, latent Meyer lemon virus and *Citrus* multiple sprouting, *Citrus* narrow leaf, etc.

Presumably, many of these viruses coexisted in their original hosts and virologists had great difficulty in isolating them one from another. Tristeza has been designated as complex, indicating the coexistence of several strains, including the Seedling yellows and the expression of symptoms depended upon the prevalence of the specific strains. Corbett and Price (1967), while conducting the studies on Psorosis, observed the coexistence of two strains of this virus along with Crinkly leaf, Infectious variegation, Concave gum, Blind pocket, and Ring-spot. Vogel and Bove (1968) found the coexistence of Cristacortis, Psorosis, and Exocortis. Tanaka and Yamada (1974) recorded the coexistence of Satsuma dwarf, Citrus mosaic, Navel infectious mottling, and Natsu daidai dwarf. Grasso (1973) observed that Ringspot, Infectious variegation and Psorosis A coexisted in lemon. Balaraman (1980) observed mixed infection of Greening, Infectious variegation mosaic, Woody gall-vein enation, and severe strain of Tristeza but showing only the symptoms of Greening and Woody gall-vein enation in acid lime. Similar coexistence has also been found with

Cachexia, Dweet mottle, Psorosis, Tatter leaf and Infectious variegation and several other cases.

Citrus Tristeza Virus (CTV)

This virus has three types of expression in orchards: (i) bud union failure on sour and bitter orange rootstock, wilting and quick decline; (ii) stem pitting; and (iii) seedling yellows. Historically, it was thought that these three types of symptom expressions were due to infection by different viruses but subsequently found to be produced by the same virus complex or strains of the same virus.

"*Tristeza*" is a Portuguese word meaning *sadness* and *melancholy*. This word was first used to refer to the mental condition of the *Citrus* growers of Brazil when they found a sharp decline in their orchards of lime, sweet orange, and mandarin on sour orange rootstocks in 1942. Subsequently, when a similar decline was found elsewhere, it was commonly called "*Tristeza*". It was formally reported from Brazil in 1949. Presumably, tristeza is an inherent component of several domesticated *Citrus* species. It started to express itself when grafting was adopted as a means of propagation. Obviously, it is a disease of the Old World, later transported to the New World. Commonly it is believed that tristeza originated in South Africa since the incidence of such symptoms has been known to occur in that country since 1900 with no specific name. It is presumed that tristeza and its vector *Toxoptera citricidus* (Kirk.) reached South Africa from India as early as 1652, may be through the ancient 'Silk route'. Tristeza and its vector were introduced in Argentina from the South Africa during the late 1920s or early 1930s. Although the incidence of tristeza was first observed in California in the 1950s, it apparently had been introduced there in budwood and trees between 1917 and the 1930s. Similarly, it may be presumed that tristeza complex has been introduced at some time or the other to all *Citrus*-growing countries of the world through importation of infected budwoods from the points of origin of different *Citrus* species, but the rate of spread varied according to the distribution of vectors and relative prevalence of other environmental conditions.

Detection of tristeza in different countries during the first half of the nineteenth century was used to be dependent on "Lime Reaction". It implies the appearance of vein clearing and stem pitting symptoms on Mexican lime or West Indian lime seedlings by grafting the budwood from the infected plants to the test seedlings. This reaction shows that besides the quick decline due to bud union failure, there may be stem pitting symptom caused by a separate component of tristeza.

Tristeza-stem Pitting

Decline of grapefruit trees caused by "stem pitting" virus in South Africa was known since 1949. Stem pitting associated with tristeza also assumed to be a problem in West Indian lime and grapefruit in Brazil in 1951. Stem pitting coupled with stunting and small-size fruit production also became a problem on the commercially important Pera sweet orange in that country. Stem pitting on grapefruit, Pera sweet orange and Mediterranean sweet orange was also reported from Argentina in 1961. Association of stem pitting and stunting in Hassaku dwarf-infected plants showed positive 'lime reaction' but differed in relation to rootstock, as Hassaku tree on trifoliate rootstock was also highly susceptible. Occurrence of stem pitting in several types of citrus in Brazil was reported in1965. It was assumed that either severe strain of stem pitting evolved with time or a new mixture of existing strains took place to make this disease so severe in many varieties of *Citrus* species. Variation in the extent of pitting also occurs in different *Citrus* varieties. Da Graca et al. (1984) reported the occurrence of a severe strain of stem pitting in South Africa. Infected grapefruit plants showed stunting, severe stem pitting, and smaller leaves with deficiency symptoms (Plate III: Figure VII.2). Stem pitting is also common on grapefruits in Florida and California but is more aggressive in California. Tsai et al. (1993) reported the occurrence of severe strains of stem pitting in Asian countries on pomelo. These reports through several decades suggest gradual changes in the status of stem pitting and its relation to various hosts throughout the world.

Severe pitting is normally found in grapefruit, grapefruit hybrids, tangelos and certain pomelo cultivars, and *Citrus macrophylla.* Certain isolates may severely pit sweet orange. Stem pitting normally is not found in sour orange, sour lemon, rough lemon, and sweet orange. But almost any seedling of any variety can be pitted by some specific strains or isolates.

Trisetza-Seedling Yellows

"Seedling yellows" of citrus was first reported from Australia. Seedling yellows virus produced tristeza-like symptoms (lime reaction) on West Indian lime but known tristeza viruses do not produce seedling yellows on Eureka lemon. Natural infection of this virus occurs in sweet orange and mandarin whereas natural infection of stem pitting was found in lemon, grapefruit and sour orange in Australia. The relationship between seedling yellows, tristeza, and stem pitting was a matter of debate for a long time as the incidence of stem pitting and seedling yellows is coupled in most of the citrus-producing countries. Wallace (1957) could obtain no seedling yellows reaction with the virus from trees of sweet orange in California that had been naturally infected with tristeza (quick decline).

According to him, this means that the naturally occurring tristeza is not associated with the virus, virus strain, or virus complex that causes seedling yellows. Wallace et al. (1965) after conducting further studies on seedling yellows contradicted the earlier views of Wallace (1957) and suggested, "symptoms of seedling yellows may result from infection with two viruses that are probably distinct and unrelated. One of them is certainly closely related to the tristeza virus. The other can be a virus having no effect on *Citrus* hosts unless combined with tristeza-like virus. This presupposes that seedling yellows symptoms result from a synergistic reaction of tristeza virus and an unidentified virus". Roistacher et al. (1979) while conducting studies on natural spread of seedling yellows and stem pitting tristeza viruses, obtained conclusive evidence of their natural spread in citrus orchards of California.

In a general way, seedling yellows do occur in nature. Distinct symptoms may be found on grapefruit, sour orange, and sour lemon. Stunting may be severe, moderate or mild. Leaves usually become smaller, chlorotic, sometimes yellow and may have pointed tips. Shoots appear to be compressed, giving a typical stunted appearance.

Complexity of Tristeza

The complexity of tristeza is apparent from the wide variability in its expression in the fields—from quick decline, dwarfing, stem pitting to yellowing of seedlings. In nature, all the components may remain together or may be separated during the process of propagation. The type of the disease normally depends on the source of the virus, composition of the complex, kind of citrus tree into which the complex is introduced, and in the case of composite trees, the kinds of *Citrus* used as scion and stock. Stem pitting can severely injure scions of grapefruit, lime, and sweet orange directly, regardless of rootstocks; seedling yellows induce stunting and yellowing at the seedling stage and tristeza causes bud-union failure of certain scions on sour orange rootstock. There also occur mild, moderate and severe strains of these three components. Gurung (1989) found mixed infection of mild, intermediate, and severe strains of tristeza in Darjeeling hills of the eastern Himalayas. In mixed infection, one strain may not become fully systemic depending on the susceptibility of the host species. It is also apparent that in one set of conditions one strain may dominate, but if the conditions are changed, another strain may dominate. If there is a subdominant severe strain in a parent tree, this may become dominant in its daughter trees under different climatic conditions.

The complexity of tristeza has been further convoluted because of its continuous genomic recombination in nature and the appearance of new or more virulent strains. From 1980 to 1991, new or severe strains of tristeza stem pitting virus were recorded in several countries notably

South Africa, Japan, Brazil, Israel, Australia, and California and Florida (USA) causing destruction to mostly oranges (Calavan et al., 1980; Miyakawa and Yamaguchi, 1981; Da-Graca et al., 1982; Roistacher, 1982; Brlansky et al., 1986; Roistacher, 1988; Bar-Joseph and Nitzan, 1991). After the standardization of the Enzyme-Linked Immunosorbent Assay (ELISA) technique to diagnose tristeza virus, it has been possible to detect symptomless hosts. A classic example is a new strain of tristeza virus that remains symptomless in Mexican lime (Bove et al., 1988).

Detection

Until 1951, incidence of tristeza disease used to be sporadically reported on the basis of symptoms and transmissibility by vectors and grafting. "Lime Reaction" was the only dependable technique available to the *Citrus* virologists until 1957. This technique is still in use as it is the only method to establish the infectious nature of the virus. There is another field-method called the "Iodine Test". This test is a very preliminary detection method based on rapid starch depletion from the feeder roots. Subsequent development of detection technology made it possible to introduce several methods those find wide application in detecting tristeza virus even at the commercial level.

Detection methods available today may be broadly classified into four groups: (i) biological, (ii) biochemical or histological, (iii) electron microscopic, and (iv) immunological.

(I) BIOLOGICAL METHODS

These methods mostly utilize "indicator plants" and "vector transmission", amongst which, the universally applied method is Lime Reaction. In this reaction, Mexican lime also known as West Indian lime, or key lime, is used as the indicator plant. When the infected budwood is grafted to the seedling of this plant, typical symptoms are produced. The symptoms range from mild leaf flecking, vein-clearing (does not persist in mature leaves), appearance of distinct dark greenish-black water-soaked areas in veins on the undersurface of the leaf, cupping of the leaf (also produced by the Vein enation virus) in case of severe strains, vein corking similar to boron deficiency (in case of Seedling yellows), and lastly stem pitting.

It is almost a confirmatory test but has many disadvantages. It requires extensive plant materials, greenhouse facilities, skilled personnel, and considerable time (minimum 4–6 weeks). Further the sensitivity of this test is very low.

(II) BIOCHEMICAL OR HISTOLOGICAL METHODS

Methods those have gained application to detection of tristeza virus are: (a) Iodine test and (b) Microscopic observation of typical parachrystalline or banded inclusion bodies or largely aggregated viral particles.

(a) *Iodine test.* Root samples (6 mm or less in diameter) are collected from the outer margin of the tree by digging them out and cutting them at an angle exposing the inner wood. A drop of iodine solution is put on the cut surface. Iodine solution is prepared by mixing 1.5 g potassium iodide with 0.3 g iodine in 100 ml water and kept in a colored bottle. This solution fails to show the typical dark blue or black reaction as it does with the cut surface of the normal roots.

(b) *Microscopic observation of inclusion bodies*: The phloem tissues of plants infected by tristeza always ccontain inclusion bodies. Garnsey et al. (1980) developed a simple staining technique to use the presence of these inclusion bodies as a diagnostic tool.

To perform this technique, freehand sections cut from fresh tissues are mounted on brass specimen holders, frozen in cryoform, and sectioned with a cryostat. Tissues may also be fixed in 3% gluteraldehyde (in 0.1 M potassium phosphate buffer, pH 7.2), before sectioning. Sections are collected in water and stained several minutes in 0.05 to 0.1% Azure A in 2-methoxyethanol and buffered with 0.2 M disodium hydrogen phosphate just prior to use. Stained sections are washed sequentially in 95% ethanol and 2-methoxyethylacetate and mounted in Euparal. Staining and washing period vary according to section thickness and stain concentration. Observing the stained sections under an ordinary microscope, inclusions are found in young phloem tissues of tristeza infected plants. Inclusions can be found in tissues from feeder roots and barks, but petioles of young expanded leaves at the abscission zone are the most reliable source for these bodies. Similar inclusion bodies may also be found in Exocortis-infected tissues and occasionally in cells of healthy tissues. But parachrystalline or banded structures, typical of tristeza inclusion bodies, are not found in them.

(III) ELECTRON MICROSCOPY

Electron microscopy to detect tristeza virus has been used since 1963. Ultrathin sections of infected tissues collected from different varieties/ cultivars of citrus species when examined under an electron microscope the association of filiform, curled, twisted particles (2000 mμ long, 10–12 mμ in diameter) with both naturally infected and inoculated plants. These could be located within 16–28 days after inoculation by vectors.

This technique has been very widely used in Israel to develop an appropriate eradication program, Bar-Joseph et al. (1974) collected barks from representative sample trees. Extracts were obtained from those barks and centrifuged for 90 minutes at 100,000 g. The resulting pellets were resuspended in 1.2 ml of 0.01 M phosphate buffer, pH 8.2, centrifuged for 10 minutes at 4,000 g and examined by an electron microscope. According to the authors' estimate, three technicians are able to prepare materials and do the diagnosis in 24 trees per day. This procedure provides rapid results but requires a costly electron microscope and ultracentrifuge. Moreover, the sample preparation for this procedure is also very laborious and its sensitivity is limited.

Garnsey et al. (1980) applied the Negative Staining Technique to simplify the procedure. Tissue is diced in several small drops of 1–2% phosphotungstic acid (PTA) dissolved in 0.1 M potassium phosphate buffer and adjusted to pH 7.0–7.2. Bovine serum albumin (0.05%) is added as a spreader. Small aliquots of stain solution are freeze-dried and reconstituted as needed. The light green extract is placed on FORMVAR-coated carbon-stabilized grids (75 × 300 or 200 mesh) for several minutes. The grids are washed several times with fresh stain and the excess stain is blotted off with filter paper. The grids are then scanned at an instrument magnification of 5,000–10,000 or more × through a 7 × binocular. This technique gives very bright pictures of the particles and less time consuming and more sensitive.

Alternatively, extracts can be prepared by dicing tissue in 0.02 M phosphate buffer, pH 7.2. The extract is then placed on a grid as above, and the grid rinsed with buffer after several minutes. The rinsed grid is then stained with 1.0% uranyl acetate. The excess is removed and grids are scanned as described above.

(IV) IMMUNOLOGICAL METHODS

The basic principles of these methods are the production of antibodies (polyclonal and monoclonal) specific to the virus and application of them in various ways to detect viral antigens. Some of the methods applied for detection of tristeza virus are, ELISA, SDS-Immunodiffusion (Garnsey et al, 1978), Fluorescence Antibody Technique (Tsuchizaki et al., 1978), Immunosorbent Electron Microscopy (ISEM), and Radioimmunosorbent Assay (Lee et al., 1981). Among these techniques, (a) ELISA using polyclonal antibodies, is the most successful one and routinely applied for indexing tristeza in all the citrus-growing countries of the world. (b) ISEM is also used for routine physical verification of the viral particles. Another technique, recently developed, is called (c) Direct Tissue Blot Immunoassay (DTBIA) for the detection of tristeza virus.

(a) Enzyme-Linked Immunosorbent Assay (ELISA) method: Bar-Joseph et al. (1980) developed a convenient method for detecting this virus at the field level using its polyclonal antibodies. This method is basically a double antibody sandwich or DAS-ELISA method. The gamma globulin is purified and conjugates between purified gamma globulin and alkaline phosphatase were prepared. Adequate standards are included in test plates to compensate for quantitative deviation in color change due to reactant concentration, incubation conditions and plate binding capacity.

Sample selection: Samples for serological tests should be rich in recently produced phloem tissues such as bark from flushes of new growth. Bark of mature limbs, young feeder roots, leaf midribs, and floral tissues can also be used. Bark of the fruit pedicel and fruit button are found to be especially good sources. Samples may be stored frozen or dried for extended periods. Tissue may be air dried or dried under vacuum and stored at –20° C to –60° C.

Efficiency and sensitivity: Usually 5–7 mg of purified gamma globulin are obtained from 1.0 ml of the antisera. Approximately 7,000 ELISA tests are possible with 6 mg gamma globulin (well volume of the plate 0.2 ml, 2 µg ml^{-1} concentration of gamma globulin for coating and 1/400 dilution of enzyme-gamma-globulin conjugate). Conjugates can be reused up to four times when infection rates are low (< 5%) and still retain 60% of their original activity. Microtiter plates can be successfully reused as many as six times by treating the plates for 60 min with 0.2 M glycine-HCl, pH 2.2 between uses.

Tristeza virus was successfully detected in major citrus cultivars including sweet orange, Eureka lemon, Mexican lime, Etrog citron, tangelos, mandarins, grapefruits, etc. In grapefruit, however, this method of testing sporadically fails, due perhaps to the host factor.

ELISA in fact is a general serological technique. It is a highly sensitive, accurate, and rapid, and flexible method for detecting any antigenic material. The major hurdles of this technique are the production of specific antisera, purification of the gamma globulin from the antisera, and preparation of the conjugates. All these methods need standardization depending on the material to be detected. As detection of tristeza is a routine exercise in citrus production programs around the world, commercial protocols are available in the international market. If a prepared ELISA kit or purified Ig G and labeled conjugates for tristeza are purchased from reliable sources and used within the specific time period, tests can be done in a very simple laboratory by an ordinary technician. The seller provides the procedural details of this test, using a commercial ELISA kit for the detection of tristeza. This test is not only suitable for detecting the virus in infected plant tissues, but it can also detect the virus in vectors.

Cambra et al. (1991) issued different immunoenzyme techniques to improve the efficiency of the ELIZA test. The authors tested 24 variations of ELISA using tristeza virus specific monoclonal and polyclonal antibodies, anti-immunoglobulins, protein A, and the biotin/avidin system. The DAS I (double antibody sandwich indirect) ELIZA variations with polyclonal antibodies used as coating antibodies and monoclonal antibodies used as intermediate antibody detected 0.1 ng ml^{-1} virus protein where the limit of detection for other variations was less than 100 ng ml^{-1} virus protein.

(b) *Immuno-sorbent Electron Microscopy (ISEM)*/Serologically Specific Electron Microscopy (SSEM): Garnsey et al. (1980) applied this technique to detect tristeza viral particles and inclusion bodies produced inside the host tissues due to its infection. Extracts are prepared by grinding approximately 200 mg young bark tissue with 1.8 ml extraction buffer (0.05 M Tris-HCl, pH 7.2, containing 0.15 M NaCl and 0.4 M sucrose). Electron microscope grids with carbon-Parlodion films are placed for 30 minutes on drops of serum diluted 1:500 with Tris buffer (0.05 M Tris-HCl, pH 7.2). Grids are then washed with Tris buffer and placed on drops of plant extracts for 1 to 4 hours. Following the reaction period, grids are washed in extraction buffer and distilled water, dipped for 30 s in 5×10^{-4} uranyl acetate (in 95% ethanol), and rinsed for 10 s in 95% ethanol. The grids are then scanned under an electron microscope.

Several modifications of this method have been devised by various workers from time to time of which the standard one has been described by Martelli (1991). The author has recommended to grinding the tissues in carborundum powder (600 mesh) or quartz sand, use of 0.3–0.5% phosphate buffer (0.1 M, pH 7.0), repeated grinding and centrifugation of the slurry (1,500–2,000 g), use of EM grid (400 mesh) coated only with carbon, coating of the grid by the antiserum diluted near or above its end point, and decoration of the virus with the diluted freshly prepared antiserum.

(c) *Direct Tissue Blot Immunoassay (DTBIA)*: Garnsey et al. (1993) developed this technique. Freshly cut stem, petiole or fruit pedicel tissue is carefully pressed to nitrocellulose membranes. The membranes are blocked by incubation in dilute bovine serum albumin and then incubated with unlabeled or biotinylated monoclonal or polyclonal antibodies. Antigen-bound biotinylated antibodies are detected by exposure to a streptavidin-alkaline phosphatase conjugate (APC) and antigen-bound unlabeled antibodies are detected by a goat anti-mouse or goat anti-rabbit Ig G APC. Localized areas of the tissue imprints of tristeza-infected plants stain intensely and are easily recognizable at 10x magnification.

According to the authors, DTBIA is rapid, requires little sample preparation, and tissue blocks can be stored at room temperature at least 30 days prior to assay. Blotted samples can be sent safely to another location for testing. This technique has been adopted for commercial diagnostic purposes.

Using polyclonal antibodies, tristeza virus can be precisely detected but these antibodies fail to separate the strains. General techniques developed so far are: (i) use of strain-specific Monoclonal Antibodies

(MAB), (ii) analysis of dsRNA, (iii) hybridization with complementary DNA (cDNA), (iv) peptide map analysis of virion coat protein, and Polymerase Chain Reaction (PCR)-based Technique.

Lin et al. (2000) developed a method for in situ immunoassay for detection of CTV sections from stems, petioles, or leaf veins fixed with 70% ethanol and incubated with specific polyclonal antiserum (PCA) 1212 or with monoclonal antibodies (MABs) MCA13, or 17G11. Bound antibodies are exposed to a substrate mixture (nitro-blue tetrazolium and 5-bromo-3-indolyl phosphate) and presence of CTV indicated by the development of a purple color visible under light microscopy in the phloem tissues of the infected plants. Severe isolate is detectable by the strain selective MAB MCA 13. The sensibility of this method is comparable to direct tissue blot immunoassay (DTBIA).

(i) *Use of strain specific MAB*: Vela et al. (1988) in Spain produced MAB specific for *Citrus* tristeza virus. Each MAB was enzyme-conjugated and used in a double antibody sandwich ELISA assay. They found that all MAB reacted uniformly against a panel of 23 strains collected from Europe, Asia, and North America. Permar and Garnsey (1988) in Florida successfully produced an MAB (MAB CTV MCA 13) that can differentiate mild and severe strains of Tristeza in Florida. Garnsey et al. (1989) made comparative studies of different Florida and Spanish MAB. These studies indicated that a number of different epitopes exist in the tristeza virus coat protein and none of the MAB described up to 1988 reacted to all the isolates of the virus.

Mei-chen and Hong-ji (1991) in Taiwan developed MAB against seedling yellows and nonseedling yellows (severe stem pitting and dwarfing on pomelo). These MAB induced three subtypes of immunoglobulins i.e, Ig G 1, Ig G 2a, and Ig M. Two of them reacted differently against mild and severe strains in double antibody sandwich indirect ELISA. They found 6 distinct reaction patterns with 13 domestic and foreign isolates tested with 2 MAB from Taiwan plus MBA MCA 13 from Florida and 3DF 1 from Spain. The MAB H 6-1 of Taiwan was distinct from other known MABs. The MAB E 3-4 reacted similar to MCA 13, but showed at least some quantitative differences with several isolates.

Kano et al. (1991) studied serological diversity of field sources of citrus tristeza virus in Japan with Spanish monoclonal antibody (3 DF 1), Florida monoclonal antibody (MCA 13), and polyclonal antisera produced in Florida and Japan. Four field collections (three mild and three intermediate) reacted strongly to 3 DF 1 and not to MCA 13. One field collection M 16 reacted to neither 3 DF 1 or MCA 13. Other field collections including mild isolates reacted strongly to both MABs. Thus the tristeza virus of Japan showed more diversity than those found in Spain and the USA as noted in the isolates of Taiwan.

Cambra et al. (1993) further studied the epitope diversity of the tristeza virus in Spain. They collected 59 isolates and detected 6 different epitopes and 7 distinct serogroups by testing against nine virus specific MBA using double antibody sandwich indirect ELISA. There observations suggest mixed infection of different serogroups in the fields

Terrada et al. (2000) produced engineered antibodies for CTV diagnosis by tissue-print enzyme-linked immunosorbent assay (ELISA) and double antibody sandwich ELISA. They determined recombinant single chain variable fragment (ScFv) that bind specifically to CTV from hybridoma cell lines 3DF1 and 3CA5.

These ScFv were genetically fused with dimerization domains as well as with alkaline phosphatase respectively. Expression of these fusion proteins in bacterial cultures produced the diagnostic reagents.

(ii) *ds RNA analysis*: This technique is based on the fact that healthy plants do not contain a detectable amount of dsRNA whereas tristeza virus-infected plants contain a dsRNA band of 13.3×10^6 kd, corresponding to the full length replicative form of the virus plus several subgenomic bands. The number and position of these bands are constant for each tristeza isolate when assayed in the same host (Roistacher and Moreno, 1991). dsRNA of tristeza virus infected plants was analyzed by several workers. Generally speaking, it is a very simple and quick method for detecting specific strains of the virus but workers encountred several limitations in field application. They reported changes in ds RNA patterns of some isolates depending on the host and season (Jarupat et al., 1988; Moreno et al., 1990). However, dsRNA patterns are not necessarily related to pathogenic capabilities of the viral isolates. Some isolates, with similar in biological characteristics, may differ in dsRNA profiles, whereas others with the same dsRNA profiles, may differ widely in biological properties (Moreno et al., 1990). Some workers found the low molecular ds RNA band ($0.5\quad 10^6$) to be associated with severe isolates inducing a seedling yellows reaction and/or stem pitting in grapefruit in one country while others found the same band in mild isolates of the virus.

(iii) *Hybridization with complementary DNA (cDNA) probes to virus genome:* This technique is based on the specific hybridization of nucleic acids with complementary strands. The viral RNA is usually attached to a membrane and the cDNA probe, labeled with ^{32}P, hybridizes on the membrane whereas the unreactive probe is washed off. Reaction is detected by autoradiography of the membrane using a special film and amplifying system (Roistacher and Moreno, 1991). Semorile et al. (1993) in Argentina, provided evidence for the usefulness of this tool in differentiating different strains of citrus tristeza virus. These authors purified ds RNA and produced cDNA by reverse transcription. cDNA was ligated to a plasmid vector and cloned in *Escherichia coli* (*E. coli*) cells. Ampicilin-resistant colonies were screened with ^{32}P labeled short copies of the cDNA. Fifty clones were isolated with inserts with 0.4 to 2.3 kbp that showed different electrophoresis patterns and after digestion with several restriction enzymes were found to differ from each other and covered most of the viral genome. The authors further conducted southern analysis by cutting all the clones using different restriction enzymes, performing electrophoresis on agarose gels, blotting and hybridizing with short copy cDNA of the dsRNA from mild and severe strains of the virus. They found that several fragments of the clones hybridized with one probe but not with the other. Extensive research, however, is necessary to make this technique field applicable. The primary requirement yet to be resolved is a wide range of highly specific probes for unequivocal classification of the virus strains.

Narvaez et al. (2000) developed a new nonisotopic procedure to differentiate tristeza virus isolates by hybridization with digoxigenin-labeled cDNA probes and different types of target RNA. Hybridization of DIG-probes with purified dsRNA or concentrated total RNA extracts spottd on nylon membranes allowed detection of CTV nucleic acid equivalent to as little as 0.1–1.0 mg infected tissue when the reaction was developed with a chemoluminescent substrate. This sensitivity was similar to or slightly better than that obtained by hybridization with a ^{32}P-labeled probe. CTV was also detectable by hybridization of DIG-probes with tissue prints from freshly cut young citrus shoots.

(iv) *Peptide map analysis of virion coat protein*: This technique consists of purifying the viral coat protein, making a partial digestion with endoproteases, and separating the resultant peptides by electrophoresis. Lee et al. (1988) studied the characteristics of the coat proteins. Guerri et al. (1990) used the peptide map of virus coat protein for identifying the strains of the citrus tisteza virus. They characterized the peptides by Western Blot using different monoclonal antibodies and distinguished tristeza virus isolates differing in biological properties and ds RNA. This method is not practical for routine field application, however. Moreover, peptide maps only reflect variability of the coat protein that may not be related to pathogenicity.

Albiach-Marti et al. (2000) developed a procedure to purify rapidly and easily a sufficient quantity of native p25 coat protein (cp) to allow comparison of CTV isolates by serological analysis of peptide maps using monoclonal and polyclonal antibodies. CTV particles were concentrated by centrifugation and purifird by agarose gel electrophoresis. The cp was extracted from gel slices enriched in virions. The purified cp was partially digested with papain-protease and the peptides generated were separated and electroblotted to a membrane. Protein-blots were tested with monoclonal antibodies and a polyclonal antibody. The serological maps generated allowed differentiation of the isolates. This method can distinguish isolates normally indistinguishable by DAS I-ELISA, dsRNA pattern and biological characterization.

(v) *Polymerase Chain Reaction (PCR) technique*: Nolasco et al. (1997) developed a diagnostic assay for citrus tristeza virus based on PCR.Viral particles are captured by solid-phase-adsorbed antibodies, and the coat protein gene is amplified by reverse transcriptional PCR. In the reagent mixture, an additional DNA probe double-labeled with fluorescent dye and a quencher is included. In the each cycle of the polymerization reaction, the probe anneals to its target strand and is cleaved by the 5′-endonuclease activity associated with Taq polymerase enzyme. At the end of the reaction, the fluorescence is measured. In healthy samples, the probe is not cleaved and the fluorescence remains unchanged at a low level.

Pathogen: Genomes and Genomic Diversity

Although failure of the scion on sour orange rootstock or quick decline has been known in most of the *Citrus*-growing countries since the introduction of grafting in citriculture, it was only in 1947 that sufficient evidence was put forward by Wallace to designate Quick decline of orange as a viral disease. Kitajima et al. (1963) demonstrated the association of filiform particles with this disease. They prepared ultrathin sections from tristeza-infected leaves of widely different varieties of *Citrus* plants and observed the sections under an electron microscope. They detected the particles both from field samples and vector inoculated plants. In inoculated plants, the presence of particles was detectable from 16 to 28 days after inoculation. The particles were flexible with helical subunits and appeared to be curled or twisted. The length of each particle was 2000 mμ and the diameter was 10–12 mμ with an internal channel of 2–3 mμ in diameter (Plate III: Figure VII.3).

Subsequent research showed the particle to be a unipartite single-stranded RNA containing low molecular weight subgenomic RNAs.

The virus is nonpersistently related to its common vector *Toxoptera citricidus* (Kirk). Under Kalimpong conditions (20 04¢N, 88 28¢E, 1209 msl), minimum acquisition and inoculation access periods were 10 and 5 min. respectively. Maximum transmission is obtained with an acquisition access period of 24 h and inoculation access period of 30 min. Pre-acquisition fasting of insects increases the efficiency of transmission (Gurung, 1989).

CTV is the first citrus virus for which the genome has been completely sequenced (19.3-kb positive-sense RNA).

Hilf et al. (1999) observed relatively consistent symmetrical distribution of nucleotide sequence in both 5′ and 3′ regions of 19.2 kb genome in cDNA sequence of CTV isolates T30 (Florida) and VT isolate (Israel). They found a dramatic decrease in sequence identity in 5′ proximal 11 kb genome sequence of T36 when compared with those observed in T30 and VT. A DNA probe derived from this region of the T36 genome hybridized to dsRNA of only 3 of 10 different Florida isolates. In contrast analogous probes from T3 and T30 hybridize differentially to the seven isolates not selected by T36 probe. Primers designed from cDNA sequence for PCR selectively amplified these 10 isolates allowing them to be classified as similar to T3, T30 or T36. According to the authors, the 5′ region of the CTV genome can serve as a measure of the extent of sequence divergence and can be used to define new groups and group members in the CTV complex.

Albiach-Marti et al. (2000a) completely sequenced the genomes from four biologically distinct isolates. These genomes are divergent in nucleotide sequence (up to 69% divergence). It is important to know whether these large sequence differences have resulted from recent evolution to design disease management strategies particularly to use genetically engineered mild strain (essentially symptomless)-cross protection and RNA-mediated transgenic resistance. The complete sequence of a mild isolate (T30), endemic in Florida for about a century, was found to be nearly identical to the genomic sequence of a mild isolate (T385) from Spain. The samples of sequences of other isolates from distant geographic locations maintained in different hosts separated with time (B252 from Taiwan, B272 from Columbia, and B354 from California) were nearly identical to the T30 sequence. The sequence differences between these isolates were within or near the range of variability of the T30 population. It is believed that the parents of these isolates have a common origin probably Asia, and that they changed during dispersion throughout the world. In other known CTV genomes the nucleotide sequence is much greater than that expected for the strains of the same virus. It may be that a high degree of revolutionary stasis may occur in some CTV population.

Lin et al. (2000a) studied the proteins of CTV with a nonbackground reaction Western blot procedure using polyclonal antibodies. Four CTV proteins (P1, P2, P3, and P4) were detected in *C. excelsa* and Mexican lime infected with CTV, using Western blot procedure with the CTV-rabbit polyclonal antibodies 1212 and 1052. They found no nonspecific background reactions in healthy or diseased plants. The patterns of specific proteins of different isolates in different hosts were disparate. They observed P1, P2, and P3 proteins in Mexican lime infected with six isolates by both the antibodies. Antibody 1052 detected a weak P4 band in blots from Mexican lime infected with the severe isolates T36, T3, and Mm2 but not by mild isolates T30, T26, and T4. In *C. excelsa* both 1212 and 1052 antibodies detected P1 in trees infected with all isolates. P2 is detected in trees infected with T16, T3, T26, T4 and Mm2 but not T30. The molecular weights of of the four proteins in most plants were about 25, 24, 21, and 18 kDa respectively. The molecular weights of P1 and P3 of T36 in *C. excelsa* were about 27 and 22 kDa respectively. The molecular weights of P1, P2, and P3 were more or less the same as those of CP, CP1, and CP2 CTV coat protein. The authors failed to identify the nature of P4.

Satyanarayan et al. (2001) developed a system to inoculate citrus plants with recombinant CTV for the production of "Pure culture" CTV in plants and for comparison of the biological properties of rCTV to that of the original parental virus populations. Full length RNAs of CTV from virions or in vitro transcribed RNA from complementary DNA, infected only a low percentage of protoplasts and produced a minimal amount of progeny virus. It was found that virions could infect 80% of the protoplasts if rCTV progeny virus was amplified by successive passage through protoplasts using virions from CTV-9-transfected protoplasts as inoculum in subsequent transfers. Progeny virions from highly infected protoplasts were slash inoculated or bombarded into plants with relatively high efficiency of infection. The rCTV retained normal functions of replication, movement, and aphid transmissibility, and had in plants a phenotype indistinguishable from that of the parental population.

Kong et al. (2000) detected and analyzed sequence variants (haplotypes) within individual CTV isolates by single strand conformation polymorphism (sscp). They estimated the genetic diversity within and between isolates by analysis of haplotype nucleotide sequences. Most CTV isolates were found to be composed of a population of genetically related variants (halotypes) among which one was predominant. High nucleotide divergence between haplophytes may also occur in certain cases.

CTV RNA contains subgenomic RNAs: Che et al. (2001) characterized three 5′ terminal subgenomic RNAs (LMT1, LMT2, LMT3) in CTV-infected cells. These RNAs were approximately 800 nts without corresponding negative strands. LMT1 and LMT2 had nodal sizes of 850 and 745 respectively. They were produced by termination during production of genomic RNAs. Occurrence of 5′ coterminal subgenomic RNAs are unique among the RNA viruses.

Molecular studies have also been extended to the understanding of the resistance found in *P. trifoliata*. Yang et al. (2001) found CTV resistance gene (Ctv) to be a single dominant gene in *P. trifoliata*. They constructed a bacterial artificial chromosome (BAC) library from an individual plant that was homozygous for Ctv and successfully cloned this gene. This finding has opened a new vista for producing CTV-resistant transgenic plants.

Spread and Control

Citrus Tristeza is an aphid-transmitted virus. Several species have been identified as vectors. These are brown *Citrus* aphid *Toxoptera citricidus* (Kirk.), melon aphid, *Aphis gossypii* (Glov.), spirea aphid, *Aphis spiraecola* Patch, black citrus aphid, *Toxoptera aurantii* (Fonsc.), *Aphis craccivora* (Koch.), *Dactynotus jaceae* L., and *Aphis citricola* Vander Goot. But not all are equally efficient in spreading the virus in the field. *T. citricidus* is the most efficient vector in the Southern Hemisphere, but not commonly found in the *Citrus* orchards of the USA. *A. gossypii* is the vector of tristeza in California whereas *A. spiraecola* and *T. aurantii* are vectors of tristeza in Florida but all three insects were found to be inefficient vectors in the early years of their detection. The rate of transmission depends upon the strain of the virus and the variety of *Citrus* serving as the source inoculum. Correlation between natural spread of the virus and the flight activity of the vector was observed in South Africa. Aphid is a seasonal insect. The build up of its population and its flight are both related to environmental conditions in general and temperature in particular. Thus natural spread of the virus is a climate and season-dependent process.

Natural spread of the tristeza virus has been studied in several countries, especially in USA, where the vector was inefficient for a long time, and in South Africa where the vector was known to be efficient from its first discovery. During the 1950s natural transmission of the virus was found in Riverside, California to be 2.4–6.3% and took 8–12 years to reach 50% natural infection of the susceptible plants. Acceleration in natural spread subsequently was recorded (Calavan et al., 1980). Seedling yellows, although imported to the USA in 1950s, remained limited in natural spread for many years. Subsequently, its spread became very extensive, surpassing that of quick decline tristeza, maybe due to mutation of the virus. High transmissibility of seedling yellows by *A. gossypii* appeared in 1964. Roistacher et al. (1980) observed changes in the transmission efficiency of *A. gossypii* in California. Its efficiency to transmit both the tristeza-quick decline and tristeza-seedling yellows reached par with the efficiency of *T. citricidus* found in other countries, albeit the efficiency differed with different isolates. Of the 23 isolates tested, 100% efficiency was found with 13, 90% with 2, and an average efficiency between 21–28 with 4; 1 isolate maintained the old transmissibility rate of 7%. The transmission efficiency of 3 isolates collected from central California varied from 18–32%, 0–91% and 100% respectively.

The natural spread drew special attention in Israel after the sudden outbreak of the disease during 1970–1973. The point of interest was the appearance of a new form of the virus. Raccah et al. (1977) found segregation of tristeza virus isolates from a single sweet orange source plant showing variation in the rates of transmission by *A. gossypii*. Similar segregation of the isolates was also recorded in Riverside, California. But all the workers were of the opinion that the donor and receptor host plants also contributed to the differences in the respective transmission efficiency. Hermoso de Mendoza et al. (1988a) found the lowest transmission efficiency ratios using lemon as both donor and receptor host, whereas highest efficiency was found with a sweet orange-sweet orange combination. In sweet orange-Mexican lime combination, the transmission efficiency of *A. gossypii* with respect to tristeza-seedling yellows isolate was 60% whereas that of tristeza-quick decline isolate was 90% in Spain. *A. citricola* in that country could transmit both the isolates but with a very low efficiency while *T. aurantii* wholly failed to transmit the virus. Interestingly, all the strains present in any donor are not transmissible by a particular vector species.

Kano and Koizumi (1991) collected 20 subcultures of tristeza by feeding *T. citricidus* on donor plants from 15 different original field sources. They also obtained 15 cultures from these original field samples. When these subcultures and cultures were tested by double sandwich indirect ELISA using two monoclonal antibodies MCA 13 and 3 DF 1, and by double sandwich ELIZA using polyclonal antibodies, the authors found 4 aphid transmitted subcultures derived from 2 field sources giving different reactions from their parent field sources. It may thus be presumed that there may be a changing dynamic state in the occurrence of strains and their transmissibility within a host system. This change is moving from a less transmissible to a more transmissible form integrated with the cross protection properties of the strains that also move from complete to incomplete protection.

Control

The primary source of the inoculum is infected budwood or infected planting materials, and which is spread in fields by aphids. Both the aspects are dealt with elsewere. Both thermotherapy and shoot tip grafting are often recommended to obtain viral-free planting materials. Tristeza symptoms are suppressed when buds are treated at high temperature (45/35° C) for 14–35 days before grafting. Hiroyuki and Yamada (1980) recommended two methods for heat treatment. One is keeping plants directly in a high temperature chamber for a long period and the other is keeping plants at a high temperature for a short period, after preconditioning at a lower temperature. The actual method of temperature treatment depends upon the heat tolerance of the concerned varieties; preconditioning at a low temperature for several weeks is considered desirable. By alternating temperatures 45/35° C for 5, 7, and 9 weeks, tristeza-free trees are generally obtained. But the practice which received wide international attention, is "Cross Protection" to minimize application of pesticides for controlling spread of the virus in fields. Cross protection is a natural phenomenon. Exploitation of this natural resource is the primary principle in raising immune plants but approaches have also been made in utilizing attenuated strains for plants immunization.

Inhibition of symptoms of severe strain of tristeza virus by inoculating *Citrus* plants with mild strains was a common phenomenon. Natural occurrence of mixtures of different strains of tristeza virus in *Citrus* was also common. After inoculating the mild strain, if it was challenged by a severe strain four months later it remained unexpressed. But the completeness of protection differed depending on the variety. Symptoms of mild strain only appeared in sour orange and Eureka lemon; intermediate symptoms (without stem pitting) with shortening of leaves and deficiency symptoms appeared in grapefruit; intermediate symptoms appeared in Galego lime (Key lime) and *Citrus webberii*, and no symptoms appeared in Pera sweet orange and Rio mandarin. The complexities in protective interference may occur in citrus tristeza virus strains. This interference may occur to be dependent on both the donor and receptor plant species. When Galego lime was preimmunized with a mild isolate from Pera sweet orange, protection did not occur. But preimmunization of Pera orange with the mild isolate from the same source exhibited protection. Preimmunization of Ruby Red grapefruit with mild isolate from Pera orange on the other hand gave no protection. Thornton et al. (1980) conducted a sustained study on cross protection of Marsh grapefruit-sour orange combination under field condition. They observed breaking of protection in some preimmunized trees after 4 years. Balaraman and Ramakrishnan (1980) isolated 6 strains of tristeza virus with a set of 9 differential hosts. Among these strains, 2 were mild and afforded a high degree of protection in different agroclimatic regions of southern India. The protection did not break during the observation period of 5 years. Koizumi and Kuhara (1984) isolated a strain from a pomelo hybrid tree (13 years old), which was ELISA negative and produced no symptom on tristeza test plants. They preimmunized Valencia orange, Marsh seedless grapefruit, and Yoshida navel orange then grafted buds from these plants on trifoliate orange or sour orange rootstocks. These were challenged with tristeza virus complex (seedling yellows, stem pitting). The authors observed a decrease of the viral content in the pre-immunized plants versus an increse in the nonimmunized and declining plants. Roistacher et al. (1988) undertook development of attenuated strains in greenhouse conditions to immunize shoot tip-grafted plants and produced several isolates effective against severe seedling yellows and stem pitting. However, they observed reversion to virulent forms with a few isolates. Chakraborty et al. (1993) tested the preimmunized field plants in southern India (Bangalore and Tirupati) using monoclonal antibodies (3 DF 1, MCA 13), and polyclonal antibodies, and applying DAS-ELISA, DAS-indirect-ELISA, and SSEM. They detected spread of severe strains in Bangalore but not in Tirupati. According to them, the limited spread of severe strains in Tirupati may be due to the limited flight activity of vectors because of the different climatic conditions in that region. Inspite of controversial reports on sustained success with cross protection of tristeza virus around the world, this technique holds much promise if the dynamic status of properly identified mild and severe strains-host-vector-climatic conditions is properly understood.

d'Orso et al. (2000) recorded uneven distribution of genomic RNA variants with infected plants. They sorted some of these variants by individual aphids and concluded that aphids probably contribute to changes observed in the CTV population following aphid transmission. Albiach-Marti et al. (2000b) confirmed such changes of genomic and defective RNAs populations of CTV isolates. They took 14 Spanish isolates and 1 Japanese isolate of CTV. These isolates were transmitted to healthy plants by aphids. Subisolates were compared with the source isolates for symptom expression and dsRNA patterns. Nine isolates showed altered dsRNA patterns but only minor variations in intensity of symptoms induced on Mexican lime. Northern blot hybridization with cDNA probes corresponding to both the 5′ and the 3′ terminals of the CTV genomic RNA showed that the dsRNA bands that could be used to discriminate between the ds RNA pattern of the source and the aphid transmitted isolates were the replicative forms of defective RNAs. Goldschmidt (2000) examined varieties Flying Dragon, Davis A, Kryder, the citranges Carrizo and Rusk, and citrandarins (*Cleopetra* Rubidoux), and Sunki English large (*C. sunki Poncirus trifoliata*) for the presence of random amplified polymorphic DNA markers in the linkage group where the CTV resistance gene was mapped in *P. trifoliata*. The closest marker gave the best results as the presence of the bands was in agreement with the resistant phenotypes. These studies suggest the scope of future production of transgenic CTV resistant plants using the resistance gene of *P. trifoliata*.

Citrus Psorosis Virus (CPV)

Psorosis is also an ancient disease of *Citrus*. It is in fact the oldest viral disease documented for these fruits. Presumably it was introduced in most of the citrus-growing countries of the world through budwood importation and subsequently spread by natural means. Etymologically Psorosis means *a state or condition of a cutaneous disease producing an itch, scabies-like symptom.* Swingle and Webber in 1896 first observed this symptom in citrus in Florida (USA) who designated the symptom by this name. In 1908, Smith and Butler recorded a similar symptom in California (USA) and designated it as Scaly Bark. Fawcett in 1933 demonstrated its viral nature or in other words, its graft transmissibility. Similar symptoms had been recorded from time to time by different names, such as Psorosis A, Psrosis B, Concave gum, Blind Pocket, Crinkly leaf, Infectious variegation, etc., all producing similar symptoms on young leaves. Because of this common feature, they were grouped as "Psorosis Complex". At one time this complex was the most destructive disease of citrus because of its natural spread and is still a very serious and destructive disease in some countries as the causal agent has yet to be physically and immunologically characterized and proper eradicative and protective measures taken up. Continuous research on this complex, cross protection and other studies conducted in various countries separated Concave gum, Infectious variegation, and Crinkle leaf from this complex. Psorosis A and B may be the same virus. Historically they are differentiated only on the basis of symptoms.

All the viruses regarded as components of the "psorosis complex" produce similar symptom in young leaves, popularly called "young leaf symptom". Typical "young leaf symptoms" produced by all these components are 'small elongated areas, white to yellowish or distinctly lighter or slightly paler green which occur on the leaves in the region of the veinlets. The veinlets as well as the tissues adjacent to or between them may show faint to pronounced clearing, often the small flecks are indistinct and fade out gradually at the margin or coalesce and form a large light green area. Sometimes a clearing extending along the veins in the form of a band is the principal feature. Frequently the cleared region forms a distinct characteristic pattern known as the zonate or oak leaf pattern. As the leaves mature, the cleared areas or young leaf symptoms disappear. On mature leaves other symptoms such as circular spots, ring spots, concentric ring spots, gum excrescence, blotches, crinkle, variegation, etc. may develop. Generally two types of symptoms, early generalized lesions consisting of faint mottling of young developing leaves and localized lesions increasing in size after several years at a very low rate are common. Local lesions occur in the bark of the trunk and limbs as more or less regularly shaped areas where the bark breaks off as scales and flakes; on the blades of leaves as oval chlorotic spots; and on the rind of fruits as multiple, circular, concentric furrows and depressions.

Psorosis A: bark symptoms are prominent in sweet orange, grapefruit, and tangerines and leaf symptoms prominent in other varieties. Severity of this virus ranges from mild to severe. Bark symptoms begin either as small scales or flakes of the outer bark with or without gum formation or an aggregation of small erumpent pustules under which the tissue is brown. These scales or pustules usually occur in localized areas on the bark of the trunk or limbs of 6–12-year-old trees. As the scaling advances, the deeper layers of barks are affected by irregular growth and by gum or gumlike deposits. Coincident with or following the development of bark lesions, gum deposits are formed within and between layers of wood more or less corresponding to the seasonal annular rings of growth. Secondary lesions are manifested by irregular discoloration of wood. Only severe strains produce leaf, twig, and fruit symptoms similar to those produced by Psorosis B. Circular spots may appear on mature leaves, lesions may appear on twigs. Fruit spots and sunken furrows or rings are rare.

Psorosis B: produces bark scaling on the trunks (stems) and larger limbs, similar to those produced by Psorosis A. But gumming is more profuse in advance of scaling. Gumming, bark splitting and scaling of twigs of all sizes are major symptoms distinguishing Psorosis A from Psorosis B. On twigs or water sprouts growing on badly diseased trunks, spots resembling those on mature leaves, with or without rings occur on the green bark. As the water sprouts mature, the raised portion of the spots becomes corky, glazed, and hard or some of the spots resemble those produced in leprosy. Primary and secondary wood symptoms are common to both A and B but their production is faster in B. Mature leaf symptoms are of different kinds, appearance from dots to large translucent areas, more or less circular and frequently in the form of rings, central portion light to yellowish-green with or without a light yellow circular to irregular border. Some spots may be in the form of single rings, concentric rings or partial rings occasionally with a tendency to necrosis in a portion of them. Another form of spots is made up of small corky raised area forming a pustule or cluster of gum-filled dots surrounded by a halo and simulating certain types of leprosy. Fruit symptoms are rings bordered by sunken grooves varying in size and pattern. On young green fruits, circular spots similar to those found on mature leaves may occur.

Concave gum: Major symptoms are the appearance of concavities of various sizes on the trunks and larger limbs. Young leaf symptom appears to be zonate or oak leaf pattern. The uniformity and regularity of appearance of such a symptom during certain growth flushes is considered a diagnostic character. Vogel and Bove (1980) found pollen transmission of this virus.

Blind pocket: The most common symptom is formation of trough-like pockets in bark and wood, eruptive with bark scaling. It produces no symptom on mature leaves.

Among these viruses physical and chemical characterization of "Infectious variegation" has been done establishing it as a completely new virus. Another virus called "*Citrus* ringspot" which often remains associated with the complex, has also been physically and chemically characterized and declared a new virus.

Detection

When sweet orange indicator seedlings are inoculated with buds from psorosis-infected *Citrus*, typical young leaf symptoms appear within 6–10 weeks. Detection of psorosis A, B, and ringspot virus may be hastened by the development of a necrotic shock reaction in sweet orange as early as 5 weeks. For other psorosis-like diseases (concave gum, blind pocket, infectious variegation and crinkly leaf), young leaf symptoms are manifested in indicator hosts 6–10 weeks postgraft inoculation.

Local lesion assay: Some strains produce local lesions in *Chenopodium quinoa, Nicotiana megalosiphon, N. benthamiana, Gomphrena globosa, Capsicum annum,* and *C. frutescens* cv Mexican chilli. First flush shoot tips are ground in ice-cold buffer (0.05 sodium borate pH 8.3 plus 0.5% mercaptoethanol) in a pre-chilled mortar and pestle, at a dilution not greater than 1:3. Inoculated leaves are washed with water shortly after inoculation. *C. quinoa* produces local lesion 7–10 days postinoculation. A few local lesions develop on *N. megalosiphon* 10–14 days post-inoculation. Spreading purple lesions develop on *G. globosa* 14–21 days postinoculation. Mild systemic mosaic develops on *N. benthamiana*. *C. frutescens* (3–6 weeks old) produces vein clearing within 12–16 weeks after inoculation; subsequently systemic vein clearing appears (Levy and Gumpf, 1991).

Levy and Gumpf (1991) also developed a technique to isolate psorosis-A specific dsRNA that can be routinely used to confirm the presence of this virus. They further developed the ELISA technique to detect the same virus using polyclonal antiserum. It was the first antiserum to a psorosis agent for its use in a diagnostic assay. Potere et al. (1999) were able to produce monoclonal antibodies and they developed an all-monoclonal-antibody based ELISA kit for the diagnosis of CPV.

D'Onghia et al. (2001) developed the direct tissue blot immunoassay (DTBIA) technique to detect CPV in infected Bonanza orange, grapefruit, lemon, and Dweet tangor. Freshly cut surfaces of different plant tissue are gently pressed on nitrocellulose membranes exposed to CPV monoclonal antibody (MAB Ps 29) conjugated with alkaline phosphatase and stained with BCIP-NBT Sigma fast.

Transmission

The psorosis comples is graft transmissible. Several workers in different countries recorded natural transmission of this complex. Some of its components are mechanically transmissible from *Citrus* to *Citrus*, *Citrus* to several herbaceous hosts and back to *Citrus*. Transmission of crinkly leaf/psorosis virus may also take place through lemon, Carrizo citrange seeds (15–31%), Troyer citrange and trifoliate orange (1–10%). Roistacher (1993) suspected the transmission through pollens. D'Onghia et al. (2000) investigated seed transmission of CPV in mandarin cv. grosso di Puglia and the Egyptian sour orange cv. Baladi but found no evidence for such transmission.

Pathogen

Levy and Gumpf (1991) conducted electron microscopic studies with a purified preparation of the virus and found it to be composed of rod shaped flexuous particles. The full length of infective particles varies from 650–665 nm and the width varies from 12–13 nm. dsRNA analysis of this virus and members of the Calra Virus Group shows close similarity.

Citrus Ringspot Virus (CRSV)

Citrus ringspot virus infects several species of *Citrus*, in particular sweet orange, mandarin, lemon, lime, and trifoliate orange.

Symptoms

After inoculation with tissue grafts, tender leaves of test plants show faint chlorotic spots. Small portions of the veins may also be cleared or chlorotic. The spots and chlorotic parts of the veins on the upper surface of the leaves later become yellowish, and green islands appear in many of the spots so that small rings are formed. Occasionally small necrotic areas develop in the spots and rings. As the leaves mature, the yellowish spots and rings become very conspicuous, and many of them coalesce to make larger blotches (Plate III: Figure VII.4). Spots, rings, leaf blotching, and vein symptoms also appear on the undersurface of the leaves but hardly any yellowing.

Sometimes, the early vein symptoms resemble the vein clearing of tristeza, but symptoms tend to spread into adjacent tissues, giving a feathery, banded appearance. Sweet orange trees occasionally display large patterns resembling mature-leaf symptoms of psorosis A.

As symptoms begin to appear, a shock effect may occur inducing leaf drop and necrosis of soft stem, or a lesion may appear farther down the stem. Occasionally, the lesion encircles the stem, causing death of the parts above.

Transmission and Detection

Garnsey and Timmer (1980) demonstrated that *Citrus* Ringspot Virus (CRSV) is mechanically transmissible from *Citrus* to *Citrus* and to a wide range of herbaceous hosts. Inoculum is prepared by grinding leaf tissue collected from recently infected plants with shock-phase symptoms, in cold 0.05 M Tris (Tris-hydroxy-methyl amino methane) buffer, pH 8.0, plus 0.5% 2-mercaptoethanol with pre-chilled mortars and pestles. Inoculum is applied with cotton swabs to leaves pre-dusted with 500-mesh carborundum. But contaminated knife and clippers fail to transmit the virus. All isolates produce characteristic chlorotic to necrotic local lesions on *Chenopodium quinoa*. Symptoms normally appear within 7–10 days after inoculation. But transmission from *C. quinoa* to *Citrus* is sporadic but consistent to *Gomphrena globosa*. Transmission from *G. globosa* to *Citrus* is consistent particularly in receptive *Citrus* hosts such as Citrog Citron. Clementine mandarin and Orlando tangelo are symptomless carriers of this virus (Vogel and Bove, 1980). Garnsey and Timmer (1988) isolated a strain which produces typical psorosis bark scaling symptoms. Rustici et al. (2000) demonstrated its mechanical transmission to different herbaceous hosts including *Phaseolus vulgaris* cv *saxa* in which it becomes systemic.

Pathogen

The pathogen has been identified as a virus on the basis of its graft transmissibility but its physical and chemical properties remained unrecognized for a long time because of the unstable nature of the infective sap. Derrick et al. (1988) improved the stability of the infectious sap by adding 0.1% ascorbic acid and 0.1% L-cysteine to the Tris medium. Utilizing this sap, they conducted sucrose density gradient ultra-centrifugation and found two different components. Mixture of these two components could induce infection but infectivity of the mixture was also found to be unstable. Gracia et al. (1991) found that a bipartite virus is associated with Psorosis/Ring Spot disease. This virus contains ssRNA as its genetic material and 50 Kd polypeptide as its structural protein(s). Byadgi et al. (1993) while working with the *Citrus* ring spot virus in India, found it to be easily transmissible by graft inoculation and dodder and not by sap, although CRSV reported from different countries were sap transmissible. They found two types of filamentous particles measuring 640 × 15 nm (designated as C) and 690 × 9 nm (designated as L) and tubules of 2,250 × 40 nm (designated as T) associated with the disease. C and L particles had nucleoprotein and contained about 6% ss-RNA whereas T particles contained only protein. They also found the association of a 29-kDa protein with both the C and T particles. These compounds have been found to have common epitopes detectable by polyclonal antisera prepared against each component. Field detection of this virus is possible by ELISA and ISEM. Different strains isolated in different countries appear to differ in protein contents. DaGraca et al. (1993, 1991),

Derrick et al. (1988), and Navas-Castillo et al. (1991) reported the presence of a 48-kDa protein in isolates from Argentina and Spain. Navas-Castillo and Moreno (1993) reported the presence of another 38-kDa protein in addition to the 48-kDa one in the Spain isolate. Gracia et al. (1991) found a 50-kDa protein in an Argentine isolate (Plate III: Figure VII.7).

According to Derrick et al. (1993), the very flexible, filamentous virus particles associated with *Citrus* ringspot and *Citrus* psorosis appear to represent a new group of plant viruses. Since the particles appear as spirals when viewed with an electron microscope, the authors tentatively named this virus group Spiroviruses. Based on serology, electron microscopy, and the presence of a 48-kd capsid protein, Spiroviruses have been associated with *Citrus* ringspot and citrus psorosis in Florida, Spain, Argentina, and Israel. These viruses have an extensive experimental herbaceous host range, but known natural infections are restricted to *Citrus*. There is considerable variation in the symptoms induced by different isolates, and serological studies using polyclonal and monoclonal antibodies indicate serological diversity among isolates.

Rustici et al. (2000) studied the Indian *Citrus* ringspot virus infecting kinnow mandarin. They purified the virus and found it to be a flexuous rod with a modal length 650 nm. They could find no relationship with Allexi-, Capillo, Potex and Trichoviruses. The purified preparation yielded a major band presumed to be coat protein of about 34-kDa and single ssRNA of 7.5 kb which was found to be infective. They proposed this virus as a new species.

4. Citrus Tatter Leaf Virus/Citrange Stunt Virus (CTLV)

Wallace and Drake in 1962 first described the *Citrus* tatterleaf virus (CTLV) on Meyer or Beijing lemon tree introduced from China to California. Afterwards this disease was reported from Florida (USA) and other countries, notably Japan, Australia, South Africa, Taiwan, and China.

It attacks mainly *Citrus* cultivars grafted on trifoliate or trifoliate hybrids causing bud union crease, yellow ring groove, yellowing and declining of canopy. In citranges (Troyer, Carrizo, Rusk, Morton) every leaf of certain shoots may be severely affected and marked distortion of the stem may be associated with irregular areas of chlorosis; corky areas may often develop on the stem. Many varieties on non-trifoliate orange rootstocks may remain symptomless carriers. Sweet orange, although symptomless is suitable for rapid multiplication.

When Meyer lemon bud wood is indexed on *Citrus excelsa*, abnormal to tatter leaf symptoms appear. Leaf blotch or psorosis-like symptoms are also observed in young and mature leaves of *C. excelsa* and also in leaves of Mexican lime. When Meyer lemon budwood is indexed on seedlings of citrange or citremon, show strong blotchy spot, deformed leaves, and stunted stems with severe pitting and grooving; hence a different name was assigned, namely "Citrange Stunt". When symptomless recovered shoots of *C. excelsa* are reinoculated with the original Meyer lemon tissue, typical tatter leaf symptoms appear in the new growth of *C. excelsa* leaves. Recovered and symptomless tissue when indexed on *C. excelsa*, showed no tatter leaf symptoms but when indexed on citrange or citremon, stunting symptoms appear. Thus tatter leaf virus remains in the apparently recovered and symptomless *C. excelsa* and the holding period may be 30–60 months, indicating that tatter leaf and citrange stunt are not caused by two different viruses but by a single virus complex (Roistacher, 1988).

Transmission
Tatter leaf is both a graft and mechanically transmissible virus. It can be mechanically transmitted to several herbaceous hosts and back to *Citrus*. Cowpea is generally regarded as the indicator plant of the vius. It could be mechanically transmitted from Etrog citron to *Nicotiana clevelandii* and back to citron.

Pathogen
Workers on the tatterleaf disease have demonstrated its graft transmissibility. Several workers also studied its physical and chemical properties but their conclusions differed. Elongated particles are found in partially purified preparations. Nishio et al. (1989) purified filamentous flexible particles 650 × 13 nm from inoculated *Chenopodium quinoa* plants using molecular permeatron chromatography on controlled pore-glass beads. They reported the virus as having single RNA molecule (molecular weight 2.83×10^6 M) and a coat protein of 27,000 Da. Since the purified particles immunoreacted with the Apple Stem Grooving Virus (ASGV), it has been suggested as a member of the Appilovirus Group (Milne, 1988). Chung and Ru-Jian (1991) observed flexible rods of variable size, 700–850 × 15 nm, 600–700 × 15 nm, 600 × 19 nm or 450–900 nm. According to them, the virus may have 2–3 components. But the infectivity of any of these particles or mixtures of them could not be consistently verified.

Marco et al. (1991), on the other hand, found icosahedral particles in cowpea and *C. quinoa* inoculated with the tatter leaf virus. dsRNA pattern analysis for these icosahedral particles nearly coincides with that of an isolate of Cucumber Mosaic Virus (CMV). They also found filamentous particles of variable length along with the icosahedral particles. According to them, the filamentous particles are not responsible for infection; the icosahedral particles are the actual viruses and hence belongs to the "Cucumovirus Group".

Hailstones et al. (2000) developed a robust and specific seminested reverse transcription polymerase chain reaction that detects CTLV in a range of *Citrus* tissues. The sensitivity of this assay is 500 times greater than that of ELISA-based methods and allows detection directly from field trees. Nucleotide sequence analysis of the amplified CTLV fragments shows near identity (99.8%) amongst Australian isolates of apple stem grooving virus (98.1%) and a high level of identity (92.0%) in a Japanese isolate of CTLV.

Citrus (Infectious) Variegation Virus (CVV)

Citrus Variegation Virus (CVV) used to be considered as a part of the *Citrus* Psorosis complex. Its graft transmissibility was regarded as diagnostic of its viral nature. After successful isolation, purification, and crystallization of the infective particles, its true viral nature was confirmed.

Symptoms

This disease is characterized by the appearance of yellow or white areas on green leaves. Psorosis-like leaf pattern may also appear as an initial symptom. Typical variegation on leaves develops within 2–4 months after infection.

Transmission and Hosts

The virus is transmitted from *Citrus* to *Citrus* by grafting. It is also transmissible from *Citrus* to *Citrus*, *Citrus* to herbaceous hosts, herbaceous hosts to herbaceous hosts and back to *Citrus* by mechanical methods. To conduct mechanical transmission, 1.0 ml of 20% sucrose and 0.05 of activated charcoal per gram of leaf tissue is used as inoculum.

Citrus is not normally an increase host for the virus. Among *Citrus*, suitable increase hosts are rooted cuttings of Eureka lemon and seedlings of Rangpur lime. Among herbaceous hosts, *Nicotiana glutinosa*, *N. tabacum* cv. Turkish, and *Phaseolus vulgaris* cv. Red Kidney are good increase hosts. In addition to these hosts, this virus can also be mechanically transmitted to cucumber, *Petunia*, and *Antirrhinum majus*.

Detection and Assay

CVV produces characteristic local lesions on cowpea and pepper. These are used for assaying the virus concentration. The plants for the assay are normally kept at 24° ± 2° C. Inocula are prepared in cold 0.05 M potassium phosphate buffer, pH 7.2, and applied with sterile cotton swabs to leaves dusted with 400-mesh carborundum. On infection, local and systemic symptoms appear in cowpea but only systemic symptoms in pepper. Antisera produced for the Florida isolate have been used for detection purposes. A serological relationship between different isolates of the virus, leaf rugose virus, and crinkle leaf has been established but not with cowpea mosaic virus. ELISA is now routinely used to detect this virus in field-grown plants (Davino and Garnsey, 1984).

Pathogen

CVV particles are icosahedrons of 25–30 nm diameter. They contain ssRNA and belong to the Ilar virus group. The genome of CVV is tripartite and consists of three single-stranded messenger RNAs, which are encapsulated separately in icosahedral particles and designated RNA-I, RNA-2, and RNA-3 according to the decreasing order of size. Two additional RNA species, designated RNA-4 and RNA-4a, are also found in viral RNA preparations. Although Citrus Leaf Rugose Mosaic Virus (CLRV) is serologically related to CVV, it does not contain RNA-4, nor do the other Ilar viruses. CVV RNA-1, RNA-2, and RNA-3 require either RNA-4 + RNA-4a fraction or coat protein to be infective. The coat proteins of other Ilar viruses can be interchanged with the coat proteins of CVV to cause a mixture of RNA-1, 2, and 3 species to be infective. Calvert et al. (1988) used the RNA of CVV to prepare a library of 76 complementary DNA (cDNA) clones. They confirmed the clones to be virus specific by hybridization analysis and also sized them by restriction digest analysis. These clones are now in use as probes to characterize the homology between various isolates.

Satsuma Dwarf Virus (SDV)

Satsuma Dwarf Virus (SDV) is an economically important disease of citrus in Japan. Yamada and Sawamura first described this disease in 1952. Its incidence has also been reported from Turkey and Korea. In China, it is limited to a few cultivars in a few locations (Zhou et al., 1993)

Symptoms

The typical symptom of this disease is the appearance of "boat-shaped leaf" in the spring flush. Some strains induce severe symptoms on sweet orange (Iwanami et al., 1991). Many citrus cultivars and their relatives may be infected with this virus and remain symptomless (Iwanami et al., 1993). China laurestine (*Viburnum odoratissimum* Ker.) commonly planted as a windbreak in Japan, becomes infected with SDV (Koizumi and Kano, 1991). According to Iwanami et al. (1993) *Swinglea glutinosa* appears to be resistant to this virus in cold temperatures (22° C day and 18° C night). No virus could be detected in this plant by the ELISA technique when grown in these low temperatures.

Transmission, Detection, and Assay

SDV is both graft and mechanically transmissible. This virus can be mechanically transmitted to sesame and 16 varieties of 9 species of leguminous plants. Studies on the mechanical transmission of the SDV have been done using leguminous and non-leguminous herbaceous plants.

When leguminous plants are inoculated with sap from Satsuma orange trees infected with SDV, striking symptoms developed in cowpea, kidney bean, and some other plants. Susceptible leguminous plants were: *Arachis hypogea* L. (peanut), *Astragalus sinicus* (Chinese milk vetch), *Canavalia gladiata* DC (sword bean), *Crotolaria spectabilis* Roth, *Kummeravia striata* (Thmb.) Schnider (Japanese clover), *Phaseolus vulgaris* L. (Satisfaction, Top crop, Nagausura, and Improved Ideal Market varieties of kidney bean), *Vigna sesquipedalis* Fruwirth (Kursanjaku and Akasanjaku varieties of asparagus bean), and *Vigna sinensis* Savi (Black-eyed and Bombay varieties of cowpea). Satisfactory kidney bean (Cashiro-ingen) developed chlorotic spots, followed by clear mottling and vein clearing on systemically infected leaves, which subsequently became malformed. Top crop kidney bean exhibited chlorotic spots, which turned into necrotic spots with concentric rings. Japanese clover showed terminal wilt. *Crotolaria* developed no apparent local lesions on infected leaves; systemically infected leaves showed vein clearing, followed by vein necrosis and soon turned yellowish and dropped. Among these susceptible plants, Black-eyed cowpea and Satisfaction kidney bean developed the most conspicuous symptoms and appear to be good indicator plants for SDV. But all attempts to transfer SDV from leguminous plants back to *Citrus* failed.

Among the 27 species of nonleguminous herbaceous plants from 12 families, only sesame (*Sesamum indicum* L.) proved to be susceptible. Several days after inoculation, the leaves of sesame developed local lesions as chlorotic spots initially that later became necrotic with a yellowish halo. The systemically infected leaves first showed vein clearing and later vein necrosis. Leaves that developed after the infection showed mottling, stunting, curling, and malformation. Cross inoculation between infected sesame and Black-eyed cowpea or Satisfaction kidney bean always gave a positive reaction, indicating the presence of SDV. Among the three sesame varieties, white sesame was found to be most susceptible. It developed very distinctive symptoms at the stage of one to two leaves.

Physalis floridiana has been found to be a good reproduction host and *Phaseolus aureus* var. Dosaka-ryokuzu a good indicator and bioassay plant for this virus.

Koizumi et al. (1988) recorded natural infection of Satsuma mandarin by SDV in fields where china laurestine was planted as windbreaks. Virus-free nursery trees became infected 2–8 months after planting in potted soil collected from SDV-infested fields suggesting soil transmission of this virus. Detection was achieved by ELISA tests.

Pathogen

Graft and mechanical transmissibility suggests the virus nature of the pathogen. Electron microscopic observations on purified preparation back in 1960s revealed the particles to be spherical with a diameter of 26 nm. But the pathogenicity and the chemical properties of these particles remain inadequately known for almost 40 years.

In 2000 Iwanami described the physical and chemical properties of this virus and its relatives. Karasev et al. (2001) observed that that the particles of SDV and two closely related viruses, *Citrus mosaic* (CiMV) and *Navel orange infectious mottling* (NIMV) are icosahedrons built of two proteins encapsulating two single-stranded genomic RNA. The genome of the SDV was completely sequenced and its replicative proteins determined. Partial genome sequence of NIMV and the phylogenetic analysis of the gene products encoded by SDV, CiMV and NIMV suggest that they may represent a new family: *plant Picorna-like* viruses. Iwanami et al. (2001) studied the sequence diversity with respect to 3'-region of RNA2 of Natsudaidai dwarf isolate (ND1) and two unidentified isolates (LB1, AZ1) and classified them into three groups as three distint virus species.

Citrus Vein Enation (CVEV)/Woody Gall

Citrus Vein Enation disease was first described in California (USA) and shown it to be caused by a virus because of its graft transmissibility. De Graca and Maharaj (1991) demonstrated the same virus caused both vein enation and woody gall diseases. VEV disease has been reported from several other countries, in particular South Africa, Peru, Japan, Spain, and more recently Brazil (Jacomino Angelo and Salibe, 1993).

Symptoms

Major symptoms of this disease are slight vein swellings, formation of small enations from the veins on the undersurface of the leaf, and knob-like galls on the stems, trunks, and roots. The reaction varies with the variation of the host and virus strain. Leaf enations may be few to many, the projections may vary in size and length, and galls may be absent to very abundant in stems and roots.

Transmission

CVEV is both graft and insect transmissible. It is transmitted by *Myzus persicae, Aphis gossypii,* and *Toxoptera citricidus.* The virus may persist in the vectors for several days. Symptoms usually appear within 5–7 days of acquisition access and 9–16 days of inoculation access time. *A. gossypii* shows a very high efficiency of transmission (90–95 %), whereas the efficiency of transmission with *M. persicae* is only 10%. Under experimental condition, *A. gossypii* may retain the virus for 2 weeks after acquisition and no transmission is found if the virus is retained for more than 21 days after acquisition (Hermoso de Mendoza et al., 1993). In *T.*

citricidus, the virus is retained for only 8 days (Maharaj and de Graca, 1989).

Detection
Several citrus species have been used as indicator plants for this virus. Mexican lime and sour orange serve as good indicators for CVEV whereas rough lemon, Volkemar lemon, Mexican lime, and Rangpur lime serve as good indicators for woody gall symptoms. Volkemar lemon normally develops galls 9 months after inoculation. When Orlando tangelo is inoculated with VEV, psorosis-like symptoms appear in the leaves and later, spoon-like leaves develop; over-sprouting may also appear and the apical buds may die.

VEV may also infect many hosts without inducing any symptom and the virus may remain undetected for several years.

Pathogen
Rogers and da Graca (1986) conducted electron microscopic studies and observed virus-like particles in the phloem of infected rough lemon plants. The concentration of these particles was found to be highest in the enations. Similar particles have also been detected in the hindgut and in the accessory salivary glands of *T. citricidus* (Maharaj and da Graca, 1988). While studying the dsRNA patterns in the infected cells, the authors observed a similarity of their patterns to those found in "Luteo viruses". Considering the common particle size (28 nm in diameter), aphid transmissibility and presence of 4 ds RNA bands, it has been suggested that CVEV may be provisionally classified as a member of the "Luteo virus Group" (da Graca and Maharaj, 1991a).

Citrus Mosaic (CiMV)

Citrus mosaic disease was first recorded in Japan in the 1940s and identified as a viral disease in 1958.

Symptoms
The general symptom of this disease is the appearance of boat-shaped leaves (in Satsuma orange) similar to that caused by Satsuma dwarf virus. The characteristic symptom is the appearance of green blotches or ring spots in the rind of fruits just developing mature coloraion.

Transmission
CiMV is normally graft transmissible, but it is mechanically transmitted to several herbaceous plants.

Detection
CiMV is detected by inoculating Goutoucheng, considered as its indicator host (Changyong et al., 1993). On inoculation yellowing, small leaf size, and leaf mosaic develop on sprouts of the spring flush. These symptoms are not modified by the presence of CTV nor are they found in seedlings infected with SDV+CTV. The virus that infects Miyamoto Satsuma and induces the formation of boat-shaped leaves gives similar reactions to that of SDV and CiMV when mechanically transmitted to sesame indicator plant or tested by ELISA (Zhou et al., 1993). Goutoucheng seedlings inoculated with SDV showed yellowing or leaf mosaic on the newly developing shoots in the spring.

Pathogen
Graft and mechanical transmissibility of CiMV suggest its viral nature. Electron microscopic studies conducted on the purified preparation showed the shape of the particles to be spherical with a dimension of 27 nm. The purified preparation produced similar symptoms in *Nicotiana rustica* that confirmed its infective nature. As the virus is serologically related to SDV in similar particles, it may be a strain of that virus.

Citrus Mosaic Badna Virus

Ahlawat et al. (1996) reported this disease from India. Symptoms of this disease in sweet orange have been known since 1975.

Symptoms
The disease is characterized by the occurrence of numerous green and yellow areas. These areas are irregular in shape and distributed all over the mature leaf. In some citrus varieties in particular pomelo and rough lemon, vein banding may also appear in leaves. The characteristic symptoms of the disease in field-infected sweet orange, and pomelo are bright yellow mottling of the leaves and yellow flecking along the veins. On experimental transmission to different varieties of citrus, symptoms appeared within 70 days after inoculation. Typical mosaic symptoms appear on pomelo but varying degrees of mosaic appeared in other cultivars. *Citrus aurantifolia* showed no symptom. In a general way, all the major commercially important citrus cultivars and the rootstocks are susceptible to this disease.

Transmission
The virus is transmitted through grafting. Experimental transmission can be done by dodder. The disease is mechanically transmissible to *Citrus decumana*.

Pathogen
Ahlawat et al. (1998) partially purified the virus and conducted electron microscopic studies. They found nonenveloped bacilliform particles in the partially purified preparation as well as in the field-infected and inoculated samples. The size of the particles was 30 × 150 nm. Reaction of the virus in immunosorbent electron microscopy (ISEM) using the antisera of common Badna viruses and PCR-mediated amplification of the viral genomic nucleic acid with Badna virus specific oligonucleotide primers, confirm that Indian Citrus Mosaic virus belongs to "Badna Virus Group".

SOME UNCONFIRMED VIRUSES

Cristacortis

"Cristacortis" has been derived from two words, 'cortex' meaning 'bark' and 'crista' meaning 'peg'. This disease was first recorded on sour orange and other citrus species.

Infection of this disease produces small depressions on the trunk, resembling those of concave gum. When the bark is removed from the concavities deep vertical pits are found in the wood with corresponding pegs on the cambial side of the bark. They can be found on the main limbs, the secondary branches and even on the young pencil size shoots. A cross section of a branch or trunk through the pit shows a line of discolored material extending from the bottom of the pit toward the center of the branch. Pits in the wood and pegs in the bark are also found in roots.

At first sight the symptoms resemble those of Blind Pocket, Concave Gum and Young Leaf symptom of Psorosis, especially the oak-leaf pattern These diseases never produce any symptom below the bud union and there are differences in the patterns of symptom expression. Cristacortis also differs from the Xyloporosis-Cachexia as its symptoms on Orlando tangelo differ in many respects from those produced by the latter.

Vogel and Bove (1972) could find no relationship of this virus with known *Citrus* viruses and considered it a new virus. In 1976, they reported the occurrence of several strains of this virus.

Impietratura Virus

The typical symptom of this disease is the formation of a lumpy rind. In Florida (USA) and Rhodesia, boron deficiency was thought to be the cause of this disease. This disease has also been reported from Greece, Israel, and Cyprus. It can infect several citrus cultivars in particular lemons, oranges, and grapefruits. The Santa Teresa strain of Femminello lemon is tolerant to this virus. Principal symptoms of this disease are the formation of very hard small fruits with gum pockets in the albedo. The pockets may be scattered over the fruit surface or localized in the calyx region. This virus can be experimentally transmitted by dodder and also by mechanical means to several herbaceous hosts. Primary local lesions or spots are formed on *Chenopodium quinoa, Gomphrena globosa, Petunia hybrida, Nicotiana tabacum* vars W. Burley and *Xanthi-nc, N. glutinosa, Datura stramonium,* and *N. rustica.* Of these plants, *G. globosa,* W. Burley tobacco, and *N. rustica* also develop similar lesion-type systemic symptoms; W. Burley tobacco develops stem necrosis, and *N. rustica* develops puckering and distortion of young leaves. When histological studies are conducted on the infected fruits of *C. volkameriana* and Ovale orange applying thionin-orange G stains secreted gum-like material in the tissue deep blue, necrotic tissue bright blue, and normal tissue yellowish orange. In the infected fruit, necrotic vascular bundles of the peel stains deep blue. In young fruits, in advanced stages of the disease, irregular cavities are found, lined by parenchymal cells surrounding small islands of normal tissue. Later, the cells lining the cavities secrete a gum-like substance that fills the cavities. In affected pedicels, vascular rays and the cambium remain normal. Some vessels of the xylem may take up blue stains. There is natural transmission of this virus and it can be eliminated by thermo-therapy at 38° C for five months.

Citrus Leprosis

Leprosis of *Citrus* has been known since the beginning of the twentieth century but it became rare after 1020s. It started to reappear from 1930s notably in South American countries—Argentina, Brazil, Paraguay, Peru, Uruguay, and Venezuela. It attacks sweet orange, citrange, citron, Cleopetra mandarin, grapefruit, lemon, mandarins, sour orange, and tangor. In earlier years, this disease was designated by different names in different countries, for example, "Florida scaly bark", "Nail head rust", Lepra Explosiva". The causal agent was suspected to be a fungus. Subsequently it was discovered that a toxin injected in the *Citrus* plant by the mite *Brevipalus oboratus* Don. feeding on leaves, fruits and stems develop leprotic symptoms. Further studies suggested that the nature of the causal agent indicated a virus and graft transmission of the causal agent was demonstrated. On inserting a patch from the lesions into mature green healthy tissues of the test seedlings, symptoms of leprosies appeared 3 weeks later and spread in some of the inoculated plants. Chagas and Rossetti (1984) successfully transmitted leprosis by inserting leaf tissues with symptoms into the bark of healthy plants, taking due precautions to avoid contamination of the mite and confirmed the graft transmissibility of the pathogen. It has now been accepted, however, that the pathogen is transmitted by the false spider mite, *B. phoenicis/oboratus.* Female mites were collected from leprotic and nonleprotic plants and inoculating healthy plants with them; plants inoculated with mites collected from leprotic sites developed symptoms within 17 days after infestation. Chagas et al. (1984) further confirmed such transmission under laboratory conditions. They obtained 48.3% transmission by the larvae of the mite versus just 8.7% transmission with nymphs and adults.

Association of short rod-like particles with the citrus leprosis disease was established as early as the 1970s but other physical and chemical nature of these particles remained unknown for a long time. Lovisolo (2001) found the leposis virus as nonenveloped *Rhabdovirus* characterized by bacilliform particles measuring 120–130 × 50–55 nm. It is mechanically transmissible to *Artiplex, Beta, Chenopodium, Gomphrena,* and *Tetragonia.*

Leaf Blotch

Galipienso et al. (2001) characterized the virus from Nagami kumquat. They observed the virions as filamentous particles (690 × 14 nm) containing a 42 kDa protein and a single-stranded RNA of about 9000 nt (Mr 3 $\times 10^6$). Infected tissues contained three species of dsRNA of Mr 6, 4.5 and 3.4 $\times 10^6$. The nucleotide sequence of several cDNA clones showed significant similarities with replication related proteins found in filamentous viruses of several genera of plants. Their studies suggest that the ssRNA is the genomic RNA of the virus, the largest dsRNA is its replicative form, and the two smaller dsRNAs may be replicative forms of 5′ coterminal subgenomic RNAs.

Vives et al. (2001) sequenced the genomic RNA (gRNA) and compared the sequence of gRNA of other viruses with similar morphology. They found the gRNA of CLBV has 8,747 nt excluding the 3′ terminal poly (A) tail. Open reading frame (ORF1) (nt 74–5,962) potentially encodes a large polypeptide of 227.4 kDa (P227), ORF2 (nt 5,962) encodes a polypeptide of 40.2 kDa (P40), and ORF3 (nt 7115–8206) encodes a polypeptide of 40.7 kDa (P41). Although the genome organization resembles that of members of the genus *Trichovirus,* considering the different biological, structural, and molecular properties, CLBV may be assigned to a new genus.

CITRUS VIROID DISEASES

"Viroid" is a virus lacking a protein coat or pathogenic low molecular weight-free nucleic acid. Ancient virus diseases such as exocortis and xyloprosis/cachexia for which the causal virus remained unknown for a long time, subsequently was found to be caused by pathogenic-free nucleic acids of definite length. Deletion and/or substitution of certain sequences occurred in the pathogenic nucleic acids in nature, developing different biological properties in them and thereby showing different types of viroids.

So far as is known, *Citrus* trees are naturally infected with five distinct viroid groups that have varying lengths of nucleic acids. These groups can be characterized by molecular size, and nucleotide sequencing, nucleic acid homology, and biological properties in various hosts, especially Etrog citron (Sano et al., 1988; Semancik and Duran-Vila, 1991; Shikata, 1990).

The five groups of citrus viroids are designated as CVd (Citrus Exocortis Viroid), CVd I, CVd II, CVd III, and CVd IV. CEV and CVd IV are comprised of single viroid grouping, whereas CVd I, CVd II, and CVd III groups each contain 2–4 viroids. CVd I grouping has 2 viroids (CVd Ia and CVd Ib) and CVd II grouping 3 viroids (CVd IIa, CVd IIb, and CVd IIc). The number of nucleotides of these viroids are: 371, 340–330, 302–297, 292–280, and 275 respectively. Among these five groups, three are variants of hop stunt viroid (HSVd), one is exocortis (CEVd), and the other xyloporosis/cachexia. CEVd and HSVd variants normally occur as mixtures in nature. The number of nucleotides and symptom intensity of different viroids on indicator hosts are given in Table 7.3.

The existence of several *Citrus* viroids in nature was discovered through sustained studies on exocortis and xyloporosis/cachexia. For proper understanding of their historical development, these two diseases are being separately dealt with along with their viroid groupings.

Table 7.3: Number of nucleotides and comparative symptom intensity of known *citrus* viroids on selected indicator hosts (Semancik and Duran-Vila, 1991)

Group	Viroid	No.of nucleotides	Symptom intensity			
			Citron	*Cucumber*	*Chrysanthemum*	*Gynura*
CEV	CEV	375	++++	+	++++	++++
I	CVd Ia	340	++	-	-	-
I	CVd Ib	330	++	-	-	-
II	CVd IIa	302	+	++	+++	-
II	CVd IIb	299	+	++	++	-
II	CVd IIc	297	+	++	++	-
III	CVd IIIa	292	+++	-	-	-
III	CVd IIIb	290	+++	-	-	-
III	CVd IIIc	285	+++	-	-	-
III	CVd IIId	280	+++	-	-	-
IV	CVd IV	275	+++	+	-	-

Citrus Exocortis Viroid (CEV) and Other Viroids

Research on Exocortis historically began in 1948 when a bark scaling disorder of trifoliate orange rootstock was found in USA, and named "Exocortis" which proved graft transmissible. This disease was found in Australia in 1949. Since then, it has been regarded as a viral disease. When the infected budwood was propagated on trifoliate orange, Rangpur lime, Troyer citrange, the grafts showed symptoms, but it took 3–4 years for their expression. As exocortis was a major disease of rootstocks, particularly of those known to be tolerant to tristeza, it became a serious challenge to virologists around the world to find a quick biological indexing method, when all attempts to purify the pathogen for electron microscopic and serological observations failed. The most confusing aspect of this disease is its wide variability of symptoms and occurrence of a large number of symptomless cultivars. This happens due to the variability of the pathogen, response of different stock-scion combinations to different pathogenic groups, age of the trees, and environmental conditions. Thus the scenario of the symptoms shows equal variability with variable combinations and conditions.

Symptoms

Early observations on symptoms were mostly made on different stock-scion combinations. In sensitive combinations, the symptoms were recorded as yellowing, splitting, shelling and gum exudation in the bark of the branches, along with reduction in height and trunk circumference. Effects on tolerant rootstocks were normally subtle. Grafts of lemon infected with exocortis produced measurable stunting of trees on tolerant rootstocks such as sweet orange, grapefruit and sour orange. Exocortis-infected Washington navel orange trees on Cleopetra mandarin and sweet orange rootstocks showed no obvious bark symptoms, but showed reduction in growth and produced less fruits. When infected sweet orange buds were budded on trifoliate orange or Rangpur lime, yellow blotching or yellow spots on the bark and bark-shelling symptoms appeared. On Rangpur lime, symptoms developed within 44 months after budding.

After discovery of the association of different groups of viroids with the exocortis, symptoms produced by different viroids in different stock-scion combinations were studied. Observed differences were used for detection or indexing purposes.

Detection

When nursery plants of Caipira sweet orange are budded with buds from infected and healthy clones of Marsh seedless grapefruit, Baianinha (navel) orange, Hamlin orange, and Pera orange, after 2 years, these plants are top worked with buds from seedlings of Rangpur lime. Within 5 months

Rangpur lime branches growing on infected plants show yellow elongated blotches and subsequently develop splits with raised borders typical of exocortis. This expression is apparently not affected by the presence of other viruses. This test was popularly known as the "Rangpur Lime Test" and long remained the only dependable method for detecting exocortis. Time taken for the appearance of symptoms in this test varies with the stock-scion combination and virulence of the pathogen. In certain cases, the time may extend up to 280 days.

Harvey lemon and Etrog citron are faster indicators than Rangpur lime. In these indicators, symptoms develop within 110–180 days. Moreover, some isolates hardly reacting with conventional rootstocks, produce distinct symptoms with these indicators. Frolich et al. (1965) utilized 'Etrog citron' to develop an indexing procedure in which leaf symptoms were found to develop within 5–10 weeks after inoculation. Roistacher et al. (1977) identified more sensitive citron selections for indexing mild exocortis. These selections are USDCS 60-13, Arizona 861, and Arizona 861-S1.

In Arizona 861-S1, the diagnostic symptoms are browning of the tips of the leaf blades, petiole wrinkle and necrosis, midnet necrosis, leaf epinasty, and stunting. Differences in the intensity of reaction could be used to differentiate the mild, moderate, and severe forms of the pathogen.

After the discovery of the different groups of citrus viroids, both molecular and biological methods were developed for their detection.

Biological Methods

Citron selections and various herbaceous hosts are used for biological indexing of different viroid groups. When extracts containing selected isolates derived from field sources are slash inoculated to *Gynura aurantiaca* DC, tomato, eggplant, *Gomphrena globosa* L., and *Datura stramonium* L., *Gynura* and tomato plants inoculated with CEVd isolates show stunting, epinasty, and leaf rugosity, characteristic of CEV infection 3–4 weeks after inoculation while other species remain symptomless. After pruning, second growth flushes show the presence of CEV in all hosts, whereas CVd II can be detected only in eggplant. When 'Suygo' cucumber is inoculated, pruned after 3 weeks, characteristic symptoms are observed on plants inoculated with CVd II containing sources. It induces darker coloration, reduction in the length of the internodes, size of the leaves with rugose leaf blades and distorted fruits. *Gynura* and cucumber are used to test the infectivity of CVd IV. CEVd can infect a large number of herbaceous hosts, CVd IV has a restricted host range, while CVd I, and CVd III have no host other than plant species belonging to Rutaceae.

Indexing on Citron: CVd I and CVd II produce diagnostic symptoms on citron. Pronounced epinasty or bending of leaves because of the point necrosis of the midribs on the undersurface of the leaf blades is observed. This symptom appears in a random fashion and affects only a few leaves of the inoculated plants. CVd IIa and CVd II b are symptomless in general on citron E 818 and E 819 but naturally occurring pure sources of CV IIa produces a mild response on Arizona 861-S 1 citron under optimal conditions of temperature and daylength. The reaction is the production of a faint necrosis of the tip of the leaf blade in a small number of leaves and sometimes a very mild petiole wrinkle and occasional petiole browning. CVd IIb produces no symptom in this plant. CVd III produces different degrees of petiole wrinkle and necrosis accompanied by midrib necrosis, causing marked leaf epinasty and some degree of mild stunting. CVd IV produces random leaf epinasty and midrib necrosis (Duran-Vila et al., 1988).

Indexing on Trifoliate orange: Roistacher et al. (1993) could biologically differentiate most of the CVds through inoculating Valencia/trifoliate plantings. They observed that CVd Ia produces a symptom called "Trifoliate deep Pit". On infection, CVd Ia induces deep pits in the wood of the trifoliate trunks with corresponding pegs in the barks. It also causes significant reduction of the trunk cross-section in Valencia scion and trifoliate rootstock and the canopy volume of the infected tree. CVd IIa induces mild to moderate bark cracking in the trifoliate trunk. The symptom produced by CVd IIb is termed "Trifoliate Finger Imprint". On inoculation, this viroid induces grooving in the trifoliate trunk resembling a squeezing or strangling of the trunk, leaving indented imprints with horizontal striations. CVd IIIb causes changes in the trunk cross-section and canopy volume similar to those as found with CVd Ia.

Molecular Method

Baksh et al. (1984) and Boccardo et al. (1984) developed detection procedures employing polyacrylamide gel electrophoresis (PAGE). The first group of authors detected CEV from citron and field-grown sweet orange and grapefruit trees. The second group applied PAGE technique to detect CEV in Clementine mandarin and Tarocco sweet orange trees grafted on trifoliate orange and sour orange respectively. Both the groups used both root and leaf samples. The second group could detect CEV in relatively crude nucleic acid and indicated further improvement of the procedure using purified nucleic acid. Baksh et al. (1984) used only purified nucleic acid. Later, Semancik and Duran-Vila (1991) illustrated the procedure for purification and characterization of viroids, which involves extraction of nucleic acid, its purification, and its sequential PAGE analysis.

Nucleic acid extraction: Tissues are collected from herbaceous and/or woody hosts. Actively growing tips are collected from the former plants 2–6 weeks, and from the latter plants 2–6 months after inoculation. Freshly collected tissues may be used or they may be powdered in liquid nitrogen and stored at –20° C for use later. Tissues are ground in a precooled extraction medium [0.4 M Tris-HCl buffer, pH 8.9 + 1% dodecyl sodium sulfate (SDS) + ethylene diaminotetra acetic acid (EDTA), 5 mM, pH 7.0, + 4% mercaptoethanol (MCE)], and phenol (water saturated and adjusted to pH 7.0, with 1 N sodium hydroxide solution) in the proportion of 1 gm:1 ml:3 ml in an icebath. The extract is centrifuged at 7,000–12,000 g for 20 min. The aqueous layer obtained after centrifugation is mixed with 3 M sodium acetate, pH 5.5 (1:10), and 95–100% ethanol (3 volumes). The mixture is kept at –20° C for 30 min. or can be stored at that temperature. It is again centrifuged at 7,000–12,000 g for 20 min. The pellets are taken out and the excess phenol removed. These are covered with Resuspension Medium (RM) containing 10 mM Tris, 10 mM potassium chloride, 0.1 mM magnesium chloride, adjusted to pH 7.4 with hydrochloric acid (1–10 ml medium/5–100 g fresh weight tissue) and agitated. The slurry is dialyzed with rapid stirring on a magnetic stirrer at 4° C overnight against 1 L of RM. The materials are then taken out and kept at 4° C for 4 h or overnight adding 4M lithium chloride. The mixture is then centrifuged at 7,000–12,000 g for 20 min. The supernatant is then separated out and kept at –20° C for 30 min. overnight after adding 3 volumes of 95–100% ethanol. The mixture is then further centrifuged at 7,000–12,000 g for 20 min. Ethanol is then drained. The pellets dried in vacuum. These purified pellets containing low molecular weight nucleic acid are resuspended in RM (100 µl/5 g fresh weight of tissue) and stored at –20° C or below. This preparation is further purified by Cellulose chromatography.

Purification of free nucleic acids: CF-11 analytical cellulose chromatography is applied to characterize different viroid RNAs by serial elution with an ethanol gradient. The aqueous sample containing the nucleic acids is made to 35% ethanol in buffer containing 0.1 M sodium chloride, 1 mM EDTA, 0.05 M Tris-HCl, pH 7.2 (STE). The solution is applied to a CF-cellulose column equilibrated with 35% ethanol-STE. Elution is done with 25% ethanol-STE and the eluant retained. The column is then washed with 25% ethanol-STE. The elution is then done with 20% ethanol-STE. The eluant is retained. The washing and elution cycle is continued with progressively reduced ethanol concentration and the ultimate elution is done only with STE buffer. The eluants are separately precipitated with 3 M sodium acetate, pH 5.5 (1/10) and three volumes of 95–100% ethanol. The precipitates are kept at –20° C for 30 min. or longer and then centrifuged at 12,000 g for 20 min. Pellets obtained are dried in vacuum and resuspended in RM buffer. Pure RNA solutions thus obtained are analyzed employing the PAGE technique.

Characterization of viroids by PAGE and molecular hybridization: Separation of viroids can be done according to their nucleotide length by conducting sequential gel electrophoresis involving migration of the sample into a standard gel (5% PAGE), followed by excision of a piece of the gel placed in contact with a second denatured gel (dPAGE) containing 8 M urea, which brings the single-stranded closed circular structure into a specific band. Manipulating the pH of the dPAGE buffers, linear and circular molecular forms of the viroids can be separated more efficiently. Bands can be detected by staining with ethidiumbromide. If staining is done, a second cycle of electrophoresis is to be conducted for collecting the bands for biological confirmation. Eluting the viroid nucleic acid from the PAGE bands, it has been possible to use it as a template to synthesize complementary DNA and a cDNA probe by random priming in presence of ^{32}P. This probe is now in routine use to detect CVds through molecular hybridizations. Nucleic acid analysis of inoculated citron by sPAGE and silver staining provides a fast and reliable method for routine indexing of all citrus viroids (Duran-Vila et al., 1993).

Imprint hybridization: Palacio et al. (1999) developed a method based on the hybridization of tissue imprints for routine indexing of citrus viroids. The overall process is extremely simple and allows quick analysis of a large number of samples. For maximum sensitivity and reliability, the inoculation of citron (Etrog citron) as a viroid amplification host is required. Hybridization against Digoxigenin-labeled RNA or DNA probes followed by detection of viroid probe hybrids using antiDIG-alkaline phosphate conjugate and chemiluminescent substrate CsPD was found suitable for detection of all viroids with the same sensivity as other available methods.

Pathogen

Exocortis was identified as a viral disease in 1949 and remained so designated until 1968 when Semancik and weathers discovered the association of infectivity of exocortis with nucleic acid preparations and confirmed in 1971 that infective free nucleic acid has unusual properties, which led them to conclude in 1972 that the pathogen of exocortis is a new species of "infectious" low molecular weight RNA in plants which Diener in 1979 termed as "viroids". Similar pathogens are also found as the causal agents of different plant diseases such as potato spindle tuber, tomato bunchy top, etc. Scoric et al. (2000) described two novel variants, CVd-S and CVd-M (mild), isolated from *Gynura aurantiaca* as a result of differences in symptom expression. The severe reaction in *G. aurantiaca* induced by CEVd-S is maintained under different conditions, but the variation in symptom expression induced by CEVd-M appears to be a response to environmental temperature.

Transmission, strains, and host reactions: Graft transmissibility of exocortis was demonstrated as early as 1949. It can be transmitted by dodder from citrus to *Petunia hybrida*, from *Petunia* to *Petunia* and back to citron by grafting, and sap from *Petunia* to 12 species of plants belonging to family Solanaceae. Contaminated grafting tools, pruning machines, and even hands may also transmit the pathogen. The pathogen remained viable for 16 h on a dry knife blade. Seed transmission may also occur in Baianinh (navel) orange trees. Exocortis pathogen may be transmitted to a wide range of herbaceous plants. Most interestingly, it infects *Gynura aurantiaca* DC and *G. sermentosa* DC, two plant species belonging to the plant family Compositae, on which symptoms appear within 14 days after inoculation. Research conducted during more than a century had led to interesting observations by various workers on different types of host reactions, suggesting the existence of a large number of strains of this pathogen. Observations on Rangpur lime rootstock trees reveal differences in the severity and type of scaling, period of incubation and degree of stunting in commercial groves and rootstock plantings. Fraser et al. (1961) in Australia found an isolate that causes stunting and not scaling on trifoliate orange rootstock. There may be grades of symptom expression on Rangpur lime and trifoliate orange rootstocks. Some isolates cause stunting without bark scaling, some isolates produce no cognizable symptom but its presence is detectable by indexing on citron. There may be three types of isolate, "mild", "moderate", and "severe" based on symptoms on Arizona 861 Etrog citron. Transmission of the "mild" strain to *Gynura* has been recorded but no symptoms in citron. Salibe (1980) observed that all inoculum sources were effective for knife inoculation to citron but not to Rangpur lime. According to him, certain citrus varieties are poorer sources of inoculum, lemon a good source of inoculum spread in nurseries, Calamondin and Tahiti lime poor hosts, and Avocado seedlings may serve as an alternative host producing yellow blotching in the area of bud union. Different types of association of different groups of CVds may cause widely divergent host reactions and symptom expressions. Recent observations on the production of pathogenesis-related proteins in certain CVd groups and host combinations (Pires Leitao et al., 1993) opens up a new dimension in the host-viroid interactions having much relevance for the management of the viroids. These authors obtained approximately 14 kDa protein in neutral extracts from Etrog citron infected with CEV. In acidic extracts from plants infected with CEVd, CVd I and CVd IV, they consistently found the presence of 23 kDa protein (PR 23).

Foissac and Duran-Vila (2000) characterized two citrus apscaviroids. Sequence variability in the PCR amplified cDNAs from the two CVds, CVd I-a and CVd IIId was analyzed. CVd III sequence was identical to previously described CVd III sequence. Important variability is encountered within the viroid population. Conversely, Cvd I-a displays population heterogeneity as shown by SSCP analysis, Sal I restriction site polymorphism, and sequences of 27 CVd 1-a cloned DNAs. The CVd1-a genomic heterogeneity is characterized by the presence of two major subpopulations with most of the divergent sequences.

Ito et al. (2001) isolated a new viroid called OS. It consists of 330–331 nucleotides having a central conserved region (CCR) characteristic of the genus Apscaviroid, showing the highest sequence similarity (68%) with citrus III viroid. This viroid by itself causes only very mild petiole necrosis and characteristically very mild leaf bending in Arizona 861-S1 Etrog citrons.

Xyloporesis/Cachexia

A disease on sweet lime producing wood pitting and brown discoloration was known as xyloporesis. There was another disease on Orlando tangelo and certain mandarins and mandarin hybrids named cachexia signified the symptoms of wood pitting and gum impregnated bark. When evaluated on indicators of xyloporosis and cachexia both diseases appeared to be caused by the same virus. Some workers preferred to keep the name cachexia as its existence was experimentally proven in 1952 whereas the experimental evidence for the existence of xyloporosis was put forward in 1961.

Both diseases are graft transmissible and expression of symptoms may take even 4 years after infection. The virus may also be mechanically transmitted. Roistacher et al. (1980) obtained ready transmission with infected knife-cut from citron to citron within 12 months after the cut.

A quick method for detection of this disease was developed. One- or 2-year old vigorous seedlings are inoculated with buds taken from the tree to be tested. Rangpur lime, Cleopetra tangerine, or any other vigorous seedling can be used. One bud from a nucellar tangelo seedling is to be budded above the inoculating buds. All sprouts are eliminated allowing only the Orlando tangelo buds to sprout. Under good growing conditions, symptoms develop within 10–24 months. Wood pitting and gummy pegs appear earlier in the most vigorous plants and in those in which the Orlando tangelo has been budded just above the inoculating bud.

The causal agent of this disease remained undetected for a long time as in the case of the exocortis. Eventually Sano et al. (1988) detected it as a viroid belonging to CVd IIb group, and is composed of 299 nucleotides

It differs from CVd IIa by 3 nucleotides deletions and 2 nucleotides substitutions The HSVd variant associated with the grapefruit dwarfing agent from Israel is similar in size to CVd IIb and shares 96% similarity in nucleotide sequences but differs in biological properties (Levy and Hadidi, 1993). In the field, it is symptomless in trifoliate orange and induces gumming, wood pitting, and bark-plugging in mandarin and tangelo. In citron, it is characterized by a latent infection.

A new viroid was identified from the cachexia disease source collection maintained at the University of California, Riverside having different electrophoretic mobility and inducing more moderate reaction on Person Special mandarin than the type citrus Cachexia. Analysis of the disease sources by sequential polyacrylamide gel electrophoresis (sPAGE) revealed a third member of CVd II group designated as CVd IIc (Semancik and Duran-Vila, 1991).

Horticultural Potentials of Viroids

One of the most desirable horticultural traits of citrus is dwarfing. The striking symptom of all viroid infections is stunting or dwarfing of plants. Cohen as early as 1968 proposed the use of CEV to obtain dwarfed trees. Subsequently, it became a concern of many citriculturists to conduct trials using suitable dwarfing sources (single or different combinations of individual viroids) aimed at identifying the scionic combinations and cultural practices to produce commercially and horticulturally acceptable densely planted dwarfed citrus groves. There are several interesting observations and cautions on the use of viroids in citriculture.

When Tahiti lime is inoculated with the mild strain of Exocortis, and observed 10–20% stunting of plants and 30–40% increase in production of out-season fruits are observed. Polizzi et al. (1991) made a field evaluation of two inoculum sources to limit tree size and biochemical analysis to determine any change in viroid content during 5 years of observation. The trees selected for this evaluation were sweet orange and clementine grafted on trifoliate orange and sour orange. Inocula used were CEV + 2 additional viroids that induce severe stunting on trifoliate orange and citrange. The authors observed more yield in Moro sweet orange grafted on trifoliate orange and citrange inoculated with both inocula. There was no modification of the pattern of the viroids during 5 years of observation. Roistacher et al. (1991) inoculated navel orange on Troyer citrange and sour orange rootstock with CEV, Ia, IIa, and IIIb viroids and observed the trees for 10 fruiting seasons. They recorded out-yield of fruits on sour orange rootstocks. Although there was initial retardation of color development, this became normal after 3 years. Fruit quality remained unchanged. Fruits from trees on Troyer citrange continued to show earlier growth and higher sugar-acid ratios. These observations suggest that ample scope remains for application of appropriate viroid/viroids for further improvement of citrus production in suitable rootstock-scion combinations.

NEMATODE DISEASES

Nematodes are microanimals. They occur in widely diversified forms and differ widely in size, shape, habits, and habitats. They have both saprophytic and parasitic properties. They normally live in soil and water as saprophytes. Pathogenic forms mostly infect animals and human beings. Several nematodes infect plants, however. These nematodes are called *phytonematodes* and are taxonomically placed in two exclusive Orders, *Tylenchida* and *Dorylaimida* belonging to an exclusive Class called "Nematoda". Although more than 180 species of plant parasitic nematodes of 39 genera have been reported from the roots of different citrus species, the pathogenic relationship has been established for only a few species, of which dominant ones are citrus nematode, burrowing nematode and lesion nematode.

Citrus Nematode (*Tylenchulus semipenetrans* Cobb)

As early as 1912, JR Hodges discovered a *Citrus* nematode occurring on *Citrus* roots in California. This nematode was scientifically identified and characterized by Cobb in 1913.

Morphology

Tylenchulus semipenetrans is an ectoparasite living on the surface or subsurface tissues of *Citrus* roots. It is medium in size with sexual dimorphism. The male is motile and the female one is mostly sedentary. Both have a narrow pointed stylet with a large basal knob. Female members variable in shape and usually protrude from the roots. Sex can be distinguished in second stage larvae. Eggs oval in shape; it is 71.4 ± 4.62 nm in length and 37.57 ± 2.19 nm in diameter. Second stage larva tapered, with slightly bent posterior end having no bursa. Total larval body length 340.83 ± 7.03 nm, and body width 14.21 ± 1.85 nm. Stylet is short, 14.21 ± 1.39 nm long; basal knob prominent and conspicuous. Length of the esophagus of second stage female larva is 95.45 ± 6.54 nm, longer than that of the senond stage male larva. Value of A and B are 25.82 ± 0.53 nm, and 3.63 ± 0.33 nm respectively (A = maximum body length/ length of body width; B = maximum body length/length of esophagus). Mature female is saccate, ventrally bent in the vulvar region anterior to short blunt tail; well-developed excretory pore is located anterior to the vulva. Value of A and B are 4.13 ± 0.59 nm and 2.89 ± 0.74 nm, respectively

(Thapa, 1991). Appearance of a typical mature female of citrus nematode is given in Plate IV: Figure VII.6.

Biology

The biology of the nematode starts at the egg stage. The uterus usually contains only one egg. Two types of larvae emerge from the egg and hatch within 12–14 days. First molt occurs before hatching. Of the two larvae, one is shorter and wider than the other. From this type of larva, a mature male develops within a week without feeding. The larger, more slender individual fails to develop without food and after feeding becomes a female nematode.

Second stage male larva usually formed before emerging from the eggmass. Stylet and the esophagus become well developed at this stage. In the third molting stage, larval movement gradually ceases and body length increases. In the fourth stage, length increases further but the esophagus degenerate and the testes become well developed. The adult male nematode emerges from this stage. Of the larvae emerging from the eggmass, usually 26% become males. Females as second stage larvae take about two weeks to locate appropriate roots for feeding and further molting. These are often found in soil samples collected from root zone, and remain alive for a short period at 20° C. When stored at 27° C, they may live for two months. On inoculating roots with these larvae, young females may appear within 21 days (Thapa, 1991).

Distribution

Byars in 1917 (cited in Byars, 1921) first established the infection of *Citrus* by the *citrus* nematode in California in seedlings of tangerines (*Citrus nobilis* Trare) and sweet orange (*Citrus sinensis* (L.) Osbeck) imported from Brazil. Subsequently, the spread of the *Citrus* nematode was observed in various citrus orchards in different *Citrus*-growing States of the USA. Concomitantly this nematode was reported from a large number of countries, confirming its worldwide distribution. It was reported from Israel, United Arab Republic, Kenya, Tanzania, South Africa, Egypt, Greece, Libya, Australia, India, Italy, Spain, Taiwan, and several other countries as well.

Symptoms

Citrus nematodes remove nutrients from the roots of *Citrus* plants and impair their normal growth and function. They may also inject toxin into the roots, affecting uptake of nutrient from the soil and normal metabolic processes. Severe infestation may cause decline or dieback. Moderate infestation causes retarded growth, production of pale yellow shorter leaves, and reduction of vigor and yield. Injuries caused by nematodes feeding on the roots facilitate invasion of several secondary pathogens, notably *Fusarium, Phytophthora*, etc.

Host Range

Baines, Clarke and Bitters (1948) reported 81 species, and varieties of *Citrus* to be susceptible to *Citrus* nematode. They also reported several Rutaceous plants that failed to produce infection by inoculation, and a few non-Rutaceous plants that did infect after inoculation. Vilardebo and Luc (1961) recorded 21 species of *Citrus*, 21 hybrids, and 11 other Rutaceous plants as hosts of *Citrus* nematode. They also identified 7 Rutaceous, and 6 non-Rutaceous plants as non-host of this pathogen. Thus the evidence is presumptive that most species and hybrids of citrus may act as the host of this nematode, while the host range among the non-Rutaceous plants is vey much limited (Cohen, 1966).

Although common rootstocks are susceptible to the *Citrus* nematode actual infection and disease development depend much upon their population in rhizospheric soil and on the root surface. Population growth of the nematodes depends upon several edaphic and biotic factors of the soil of the concerned locality. So the natural incidence of the nematode widely varies from location to location and season to season. Such variation was recorded during a study conducted in mandarin orchards in two agroecologically different locations of Darjeeling hills in two different seasons. Susceptibility of different citrus rootstock species was also tested by soil inoculation at different doses of nematode population to determine the threshold level for infection. Rootstock species tested were Rangpur lime, citron, Rough lemon, grapefruit, Kagzhi lime, sour orange, Eureka lemon, Cleopetra mandarin, trifoliate orange, and mandarin (Darjeeling orange). All the species were found to be susceptible excluding trifoliate orange. 100 nematode/250 g soil initiate infection in two-month-old test seedlings within six months after inoculation (Thapa, 1991).

Management

Management of nematodes in soil is a very complex process that must be integrated one beginning from the nursery stage to orchard protection. Management commences application of pesticides in the infested soil before planting as a preventive measure, or at the postplanting stage as a curative measure. Soil fumigation was commonly practiced as a preventive measure, and treatment of soil by nonfumigative chemicals was common for curative measures. But some countries also use fumigants with irrigation water at the postplanting stage. Common fumigants and other chemicals used to eliminate citrus nematodes in various countries and the respective doses of their application as a preventive measure are methyl bromide (MB) 100–200 g m^{-2} 50 kg ha^{-1} (a.i), MIT 410 L ha^{-1} (a.i),

1,3-D 600–1000 L ha^{-1}, or phenamiphos 50 kg ha^{-1} (a.i), and EDB (83% a.i) 400 L ha^{-1}. Pesticides used as preplant fumigation are MB at the rate of 50 ml m^{-2}, EDB 110 ml m^{-2}, MIT 100 ml m^{-2}. Postplanting are DBCP 15–45 L ha^{-1} with irrigation water. This treatment not only reduces the nematode population, but also increases yield by 20%. Other chemicals used at the postplanting stage are oxamyl, aldicarb, miral, ethoprop, phensulfothion and phenamiphos, at the rate of 2.4 kg ha^{-1} (a.i) (Lo Guicide, 1981).

Van Gundy and Garabedian (1981) did a comparative study on the efficacy of several fumigant and nonfumigant nematicides for the control of *Citrus* nematodes on oranges, grapefruit, and lemon. Chemicals tested were aldicarb (15 G, and 3 L) 5.6–11.2 kg (a.i) ha^{-1}, phenamiphos (15 G, and 3 L) 16.8 kg (a.i) ha^{-1}, oxamyl (L) 6.8 kg (a.i) ha^{-1}, 1,3-D 37.4–187 L ha^{-1}, EDB at the rate of 23.4, and 46.7 (a.i) L ha^{-1}. Methods of postplant application included broadcast, injection on one side of the tree in alternate years, injection or incorporation in the irrigation furrow, spaghetti-drip irrigation systems, and flood irrigation. The EDB, and 1,3-D fumigant gave good nematode control but were phytotoxic to roots in the treated area at all levels. Fruit yield and fruit size were reduced after the first year. Aldicarb and oxamyl reduced nematode numbers and increased root growth, fruit size, and yield after the first year. Phenamiphos reduced nematode numbers and increased root growth but did not increase yields after the first year. The various fumigants and nonfumigants used by different workers in varius countries are given in the Table 7.4.

Lo Giucide (1981) observed several cultural practices for nematode management. Trifoliate orange would be the first choice as rootstock for developing orchards in nematode-infested soil. In the case of a new plantation, it is to be raised in coarse sandy loam, as nematodes do not grow well in this type of soil. Disease does not develop if the poatassium supply is increased. Application of 300 g N $tree^{-1}$ (ammonium nitrate/ ammonium sulfate) each year normally eliminates nematodes.

Thapa (1991) did a comparative study on rejuvenation of Darjeeling orange plants, declining due to severe nematode infestation, by applying pesticides (different doses of carbofuran 3G, and phorate 10 G, phenamiphos 10 G), organic maure (neem cake), and mixed cropping with allelopathic plants (allelopathic relationship was previously tested by artificial trials). These trials indicated the beneficial use of neem cake, and allelopathic crops. Use of chemicals gave inconsistent results. Application of neem cake at the rate of 1500 g $tree^{-1}$, and planting marigold and cabbage in the rhizosperic zone gave promising results.

Application of natural antagonists is a very effective and environmentally safe method for controlling nematodes in orchards. Natural microbial antagonists include predatory nematodes, fungi, and

Table 7.4: The rates of fumigant and nonfumigant nematicides used in different experiments conducted in different parts of the world

Common name	Rate used	Reference
Fumigant		
MB	50 kg ha^{-1}	Ashkenazi and Oren, 1977
	300, 600, 900, 1250 kg ha^{-1}	Basile et al., 1977
EDB	100 L ha^{-1}	Ashkenazin and Oren, 1977
DBCP	18.7, 30, 42.1 L ha^{-1}	Kalyviotis-Gazelas and Koliopanos, 1972
	45, 90 L ha^{-1}	Koliopanos et al., 1979
	2.4, 3.6 kg ha^{-1}	Badra and Elgindi, 1979
	20, 40, 80 kg ha^{-1}	Elgindi et al., 1976
	30, 54.5, 70 L ha^{-1}	Koliopanos et al., 1976
	60, 90 kg ha^{-1}	Velardebo et al., 1975
MIT	1000 t ha^{-1}	Ashkenazin and Oran, 1977
Nonfumigant		
Ethoprop	30, 60 L ha^{-1}	Koliopanos et al., 1979
	2.4, 3.6 kg ha^{-1}, 30, 60 kg ha^{-1}	Velardebo et al., 1975
	10 kg ha^{-1}	Vovlas et al., 1973
Phensulfothion	15, 30 kg ha^{-1}	Koliopanos et al., 1976
	2.4, 3.6 kg ha^{-1}	Badra and Elgindi, 1979
	20, 40, 80 kg ha^{-1}	Elgindi et al., 1976
Phenamiphos	30, 60 kg ha^{-1}	Vilardebo et al., 1975
	2.4, 3.6 kg ha^{-1}	Badra and Elgindi, 1979
	50 kg ha^{-1}	Vovlas et al., 1973
Thionazin	2, 5, 10 L ha^{-1}	Kalyviotis-Gazelas and Koliopanos, 1972
	10 kg ha^{-1}	Vovlas et al., 1973
Aldicarb	2.4, 3.6 kg ha^{-1}	Badra and Elgindi, 1979
	20, 40, 80 kg ha^{-1}	Elgindi et al., 1976
	10 kg ha^{-1}	Vovlas et al., 1973
Oxamyl	2.4, 3.6 kg ha^{-1}	Badra and Elgindi, 1979
	20, 40, 80 kg ha $^{-1}$	Elgindi et al., 1976
	4, 8, 16, 32, 64 kg ha^{-1}	Vovlas et al., 1973

Source: Lo Giudice 1981

bacteria (Mankau, 1980) and have been reported from various parts of the world (Gasperd and Mankau, 1986; Fattah et al., 1989; Walter and Kaplan, 1990). Roccuzzo et al. (1992) studied the occurrence of microbial antagonists naturally balancing the *Citrus* nematode population in soil. They isolated several nematophagous fungi from different parts of Italy (Table 7.5). These fungi may be profitably used as bio-control agents in due course of time.

Proper integrated management is to start from the beginning of the production system. Seeds for the nursery should be free from any contamination. Nursery beds and soil of the Foundation Blocks for producing grafted planting materials should also be made nematode-free. Sample indexing is to done with the planting materials before selling or sending for transplantation in orchards. If moderate infestation is found

Table 7.5: Some nematophagous fungi and their mode of action

	Fungus	Mode of action
1.	*Arthrobotrys oligospora* Drechler	Adhesive network
2.	*A. conoides* Fresenius	Adhesive network
3.	*A. dactyloides* Drechler	Constricting rings
4.	*Monacrosporium salinum* Cooke et Dickinson	Adhesive network
5.	*M. elipsosporium* (Grove) Cooke et Dickinson	Adhesive knob
6.	*Dactylella bembicodes* Drechler	Constricting rings
7.	*D. acrochaeta* Drechler	Constricting rings
8.	*Trichotecium cytosporium* Duddington	Adhesive network
9.	*Catenaria anguillulae* Sorokin	Endoparasite
10.	*Meria coniospora* Drechler	Endoparasite
11.	*Verticillium chlamydosporium* (Drech.) Dow et al.	Egg parasite

Source: Roccuzzo and Ciancio, 1992

on the roots of the seedlings/planting materials, treatments are undertaken to free them of nematodes by dipping the roots in Thimet 10 G (2,000 ppm) for 30 min. then rinsing the roots thoroughly in water (Thapa and Mukhopadhyay, 1994). Appropriate biocontrol agents may be applied in orchard soil for natural control of the nematode population.

Burrowing Nematodes (*Radopholus similis* Cobb 1893)

Radopholus similis also belongs to the Order Tilenchida, but placed under the family Pratylenchydae. It is a migratory ectoparasite, and shows sexual dimophism. Adult males are noninfective. Females lay eggs in the infested tissues. Hatching takes place within 8–10 days to produce larvae maturing within 10–13 days. Completion of its life cycle (egg to egg) usually takes 20–25 days. Characteristic symptoms of reddish brown lesions throughout the cortex of the root usually develop 3–4 weeks after infection.

Burrowing nematode has a very wide host range. Poucher (1967) recorded 244 hosts of this nematode. It is distinguished into two Biotypes: *Banana*, and *Citrus*. Banana Biotype does not infect citrus. *Citrus* biotype is designated as a separate species named *Radopholus citrophyllus* Huettel, Dickson et Kaplan.

This nematode spreads by mechanical means, through planting materials, planter's hands, shoes, clothes, planting appliances, irrigation streams, contaminated soil, etc. It may survive 12 weeks outside its hosts.

Lesion Nematode (*Pratylenchus coffeae*)
(Citrus slum disease of Florida)

Pratylenchus coffeae belongs to the Family Pratylenchydae. It is a migratory endoparasite in root cortex where it feeds, and multiplies. Completion of its life cycle usually takes 45–48 days. It is temperature-sensitive. It does not surive above 38° C. Optimum temperature for its growth and development is 29.5° C

Lesion nematode produces brownish lesion on the root. Severe infestation may cause poor growth of the plant, under-sized fruits, and even dieback. Normally its infestation facilitates invasion by secondary soil-borne pathogens. *Citrus limon, Citrus sinensis, Citrus reticulata,* and banana are very susceptible to this nematode. Trifoliate orange, and *Microcitrus* are resistant to it.

REFERENCES

A. Fungal Diseases

Albrigo LG, and Young RH (1981). Phloem zinc accumulation in *Citrus* trees affected with blight. *Hort. Science* 16 (2): 158–160.

Alcain JC de, and Marmelicz LA (1984).*Ceratocystis fimbriata* Ellis, and Hulst: A new lemon tree pathogen. *Proc. Int. Soc. Citriculture,* vol. 2, pp. 432–435.

Anonymous (1997). NRC for *Citrus*: Perspective Plan. Indian Council of Agricultural Research (ICAR), Nagpur.

Beretta MJG, Pompeu J Jr, Derrick KS, Lee RF, Barthe GA, and Hewitt BG (1992). Evaluation of rootstocks in Brazil for resistance to declinio. *Proc. Int. Soc. Citriculture* Vol. 2, pp. 841–843.

Cacciola SO, Pane A, Perrotta G, and Petrone G (1992). Polygalacturas activity of *Phoma tracheiphila. Proc. Int. Soc. Citriculture,* vol. 2, pp. 884–886.

Cacciola SO, Pane A, Parisi A, Tringali C, and Pennisi A (1990).Growth curves of *Phoma tracheiphila* (Petri) Kanc. et Ghik. On liquid media and toxicity of culture fluids. *Aspettichimicie fisiologicidelle fitotossine,* Viterbo 13–15 Septembre.

Carlos EF, Lemos EMG, and Donadio LC (2000). Citrus Blight. *Laranja* 21 (1): 175–203.

Davino M, Gentile F, and Gamberini O (1992). Effectiveness of Fosetyl-Al against citrus brown rot, and green mold. *Proc. Int. Soc. Citriculture,* vol. 2, pp. 859–861.

Deng Ziniu, Gentile A, Malfa S la, Domina F, Tribulato E, Deng ZN, and la Malfa S (2001). Innovative techniques for the identification of genes in *Citrus*. Rivista di Fruticultura e di Ortofloricultura 63 (2): 45–48.

Derrick KS, Barthe GA, Hewitt BG, and Lee RF (1990a). Proteins associated with *Citrus* blight. *Plant Dis.* 74: 168–179.

Derrick KS, Barthe GA, Hewitt BG, and Lee RF (1990b). A spot test for *Citrus* blight. *Citrus Ind.* 71 (2): 56–57.

Dutta Roy S (1994). Situation of powdery mildew disease on mandarin orange (*Citrus reticulata* Blanco) in Darjeeling district. *Indian Agric.* 38 (1): 55–59.

Dutta Roy S, Mukherjee N, and Mukhopadhyay S (1991). Standardization of inoculation method and screening of different *Citrus* species against powdery mildew disease. *J. Mycopathol. Res.* 29 (2): 149–154.

Dutta Roy S, Mukherjee N, and Mukhopadhyay S (1994). Powdery mildew of mandarin orange in Darjeeling district: site of initiation, progress and perennation. *Indian J. Mycol, and Pl. Pathol.* 24 (1): 53–54.

Fagoaga C, Rodrigo I, Conejero V, Hinarejo SC, Tuset JJ, Arnau J, Pina JA et al. (2001). Increased tolerance to Phytophthora citropthora in transgenic orange plants constitutively expressing a tomato pathogenesis related protein PR5. *Molecular Breeding* 7 (1): 175–203.

Feichtenberger E, Rossetti V, Pompeu J Jr, Sorbinho JT, and de Figueiredo (1992). Evaluation of tolerance to Phytophthora species in scion rootstock combinations of *Citrus* in Brazil—A review. *Proc. Int. Soc. Citriculture*, vol. 2, pp. 854–858.

Ghini P, Marques JF, Tokunaga T, and Bueno SCS (2000). Control of *Phytophthora* sp. and economic evaluation of a solar collector for substrate disinfection. *Fitopatologia Venezolana* 13 (1): 11–14.

Ippolito A, Ciccio V De, Cutuli G, and Salerno M (1987). The role of infected *Citrus* fruits and seeds in the spread of mal secco disease. *Proc.7th Congr. Medit. Phytopath. Union*, Grenada, Spain, pp. 166–167.

Kuhara S (1999). The application of the epidemiological simulation model "MELAN" to control citrus melanose caused by *Diaporthe citri* (Faw.) Wolf. Extension Bulletin: Food and Fertilizer Technology Center. 481 (11): 6.

Lio GM di San, Tuttobene R, and Persotta G (1984). Inoculum density of *Phytophthora* species and disease incidence in citrus orchards and nurseries in Brazil. *Proc. Int. Soc. Citriculture*, vol. 1, pp. 334–338.

Lio GM di San, Cacciola SO, Pane A, and Grasso S (1992). Relationship between xylem colonization and symptom expression in mal secco infected sour orange seedlings. *Proc. Int. Soc. Citriculture*, vol. *2*, pp. 873–876.

Magnano dí San Lio G, Cacciola SO, Pane A and Grassos (1992). Relationship between xylem colonization and symptom expression in Mal Secco infected sour orange seedlings. *Proc. Int. Soc. Citriculture* Vol. 3: 873–876.

Menge JA (1989). Dry root rot. *In*: Compendium of Citrus Diseases (eds. JO Whiteside, S M Garnsey, and LW Timmer). APS Press, St Paul, MN. (USA). pp. 14–15.

Natoli M, Petrone G, Cacciola SO, and Pane A (1990). Characterization and phytotoxicity of pectic enzymes produced by *Phoma tracheiphila* (Petri) Kanc.et Ghik. *Aspettichimicie fisiogicidelle fitotossine*, Viterbo 13–15 Septembre.

Nemec S (1984). Characteristics of *Fusarium solani* infected pioneer roots on blight diseased and healthy *Citrus. Proc. Soil, and Crop Sci. Fla* 43: 177–183.

Nemec S (1985). *Citrus* blight in Florida *In*: Review of Tropical Plant Pathology (eds. SP Raychaudhuri and JP Verma), vol. 2, pp. 1–28.

Nemec S and Oswald W (1991). Cuticle and wall degrading enzymes of *Fusarium solani* and *Phytophthora parasitica* from citrus tissues and soils. *Phytopathol.* 81: 1196.

Nemec S and Baker R (1992). Observations on *Fusarium solani* Naphthazarin toxins, their action, and potential role in citrus plant diseases. *Proc. Int. Soc. Citriculture,* vol. 2, pp. 832–837.

Nemec S, Achor DS, and Albrigo G (1986). Microscopy of *Fusarium solani* infected rough lemon *Citrus* fibrous roots. *Can. J. Bot.* 64: 2840–2847

Nemec S, Baker RA, and Tatum JH (1988). Toxicity of dihydrofusarubin and isomarticin from *Fusarium solani* to *Citrus* seedlings. *Soil Biol. Biochem.* 20: 493–499.

Nemec S, Jabaji-Hare S, and Carest PM (1991). ELISA, and immunological detection of *Fusarium solani* produced naphthazarin toxins in *Citrus* trees in Florida. *Phytopathol.* 81: 1497–1503.

Parisi A, Tringali C, Lio GM di San, and Cacciola SO (1992). Phytotoxic activity of mellein, a low molecular weight metabolite of *Phoma tracheiphila. Proc. Int. Soc. Citriculture,* vol. 2, pp. 884–886.

Polizzi G and Azzaro A (1992). *Botrytis* blight in citrus nursery in southern Italy. *Proc. Int. Soc. Citriculture,* vol. 2, pp. 848–850.

Polizzi G, Lio GM di San and Catara A (1992). Dry rot of citranges in Italy. *Proc. Int. Soc. Citriculture,* vol. 2, pp. 890–893.

Rensberg JCJ Van, Labuschagne N, Nemec S and van Rensberg JCJ (2001). Occurrence of *Fusarium* produced naphthazarins in citrus trees and sensitivity of rootstocks to isomarticin in relation to citrus blight. *Plant Pathology* 50 (2): 258–265.

Thanassoulopoulos CC and Manos BD (1992). Current status prognosis and loss assessment of mal secco (*Phoma tracheiphila*) of citrus in Greece. *Proc. Int. Soc. Citriculture,* vol. 2, pp. 869–871.

Whiteside JO (1984). Infection of sweet orange fruit in Florida by a common biotype of *Elsinoe fawcettii. Proc. Int. Soc. Citriculture,* vol. 1, pp. 343–346.

B. Bacterial Diseases

Almeida AG, Sayama K and Tauyumu S (2000). Isolation of a protein bound to canker-forming factor from citrus plant. *J. Gen. Plant Path.* 66 (2): 138–143.

Aubert B (1984). The Asian and African citrus psyllids *Diaphorina citri* Kuwayma, *Trioza erytreae* (Del Guecio) [Homoptera: Psyllidae] in southwest and Saudi Arabia. Proposal for an integrated control program: Report to FAO, Rome.

Aubert B (1985). Populations de *Diaphorina citri* observees dans les vergers agrumicoles bresiliens (Est de Sao Paulo, Bahia, Sergipe) en l'absence de symptoms de greening. Document IRFA.

Aubert B (1986a). Rapport d' expertise concemant les vergers d' okoloville en Republique du Gabon, SOSUHO/IRFA-CIRADAubert B (1986b). Problems poses a l'agrumiculture Camerounaise. Rapport IRA Cameroun IFRA-CIRAD.

Aubert B (1987). *Trioza erytreae* Del Guercio and *Diaphornia citri* Kuwayama (Homoptera: Psyllidae), the two vectors of citrus greening disease: biological aspects and possible control strategies. *Fruits* 42: 149–162.

Aubert B (1992). *Citrus* greening disease, a serious limiting factor for citriculture in Asia and Africa. *Proc. Int. Soc. Citriculture,* vol. 2, pp. 817–820.

Aubert B and Quilici S (1984). Biological control of the African and Asian citrus psyllids (Homoptera: Psyllidae) through eulophid encyrtid parasites (Hymenoptera: Chalcidoidea) in Reunion island. *Proc. 9th Conf. IOCV* (eds. LW Timmer, and JA Dodds), Univ. Calif. Riverside CA, pp. 100–108.

Aubert B, and Vullin G (1998). *Citrus* nurseries and planting techniques GTZ/CIRAD: 155–157.

Aubert B, Garnier M, Guillamum D, Herbagyandono B, Setiobudi, and Nurhadi F (1985). Greening, a serious threat for the citrus production of the Indonesian Archipelago: future prospects of integrated control. *Fruits* 40: 549–563.

Aubert B, Garnier MG, Cassin JC, and Bertin Y (1988). Citrus greening disease survey in east and west African countries, south of Sahara. *Proc. 10th Conf. IOCV* (eds. LW Timmer, SM Garnsey and L Navarro) Univ. Calif. Riverside CA, pp. 231–237.

Baoqing Fang, Qian Jumei, and Luo Luyi (1991). New progress in examining and testing methods and controlling technology for citrus canker in China. *In:* Rehabilitation of citrus industry in the Asia and Pacific Region (eds. Ke Chung and SB Osman). UNDP Regional Project RAS/86/022: 192–195.

Becu P and Whittle A (1988). The Indonesian citrus variety improvement program—a costing study. FAO Field Document Project INS/84/007, Rome.

Bertha MJG, Lee RF, Barthe GA, Thome Neto J, Derrick KS and Davis CI (1993). *Citrus* variegated chlorosis: detection of *Xylella fastidiosa* in symptomless tree. *Proc. 12th Conf. IOCV* (eds. P Moreno, JV da Graca and LW Timmer). Univ. Calif, Riverside CA, pp. 306–310.

Bertha MJG, Harakawa R, Chagas CM, Derrick KS, Barthe GA, Ceccard TL, Lee RF (1996). First report of *Xylella fastidiosa* in coffee. *Plant Disease* 80: 21 (abstract).

Bove JM, Garnier M, Ahlwat YS, Chakraborty NK, and Varma A (1993). Detection of the Asian strains of the greening BLO by DNA–DNA hybridization in Indian orchard trees and Malaysian *Diaphorina citri* psyllids. *Proc. 12th Conf. IOCV* (eds. P Moreno, JV da Graca, and LW Timmer), Univ. Calif. Riverside CA, pp. 258–263.

Catling HD (1968). Report to the Government of the Philippines on the distribution and biology of *Diaphorina citri,* the insect vector of leaf mottling (greening) disease of citrus. FAO/TA Report 2589, FAO, Rome.

Catling HD (1969). The binomics of the South African citrus psyllid *Trioza erytreae* Del Guercio, I. The influence of flushing rythms of citrus, and factors that regulate flushing. *Journal of Entomological Society South Africa* 32: 191–208.

Chagas CM, Rossetti V, Beretta and MJG (1992). Electron microscopy studies of a xylem-limited bacterium in sweet orange affected with *Citrus* variegated chlorosis disease in Brazil. *J. Phytopathol.* 134: 306–312.

Chao HY, Chiang SL, Lee SL, Chiu CS, and Su WP (1979). A preliminary study on the relation between the prevalence of the *Citrus* yellow shoot (Huang glung bin), and the citrus psyllid *Diaphorina citri* Kuwayama. *Acta Phytopathologica sinica* 9: 121–126.

Chatisathian J and Tontyaporn S (1990). *In vitro* culture for the production of disease—free *Citrus* in Thailand. *In*: Rehabilitation of Citrus Industry in the Asia and Pacific Region. UNDP-FAO Regional RAS/86/022 (eds. B Aubert, S Tontyaporn, and D Buangsuwon), pp. 69–76.

Chen P (1943). Research report on citrus yellow shoot disease in Chouchow, and Santou areas. *New Agriculture* 3: 143–177.

Chen TS (1991). A scheme for eradicating citrus huan glung bin-greening in Wenzhou. 6th Asia Pacific UNDP-FAO Conf. on Integrated Citrus Health management (eds. C Ke and MO Shamshudin), pp. 36–39.

Chang CJ, Garnier M, Zreik L, Rossetti V, and Bove JM (1993a). Culture and serological detection of the xylem-limited bacterium causing *Citrus* variegated chlorosis and its identification as a strain of *Xylellia fastidiosa. Current Microbiol.* 27: 137–142.

Chang CJ, Garnier M, Zreik L, Rossetti V, and Bove JM (1993b). *Citrus* variegated chlorosis: cultivation of the causal bacterium and experimental reproduction of the disease. *Proc. 12th Conf. IOCV* (eds. P Moreno, JV da Graca and LW Timmer), Univ. Calif. Riverside CA, pp. 294–300.

Chung Ke (1991). The present status of citrus huan glung bin and its control in China. Rehabilitation of *Citrus* industry in the Asia and Pacific Region (eds. Ke Chung, M Shamshudin, and B Osman). UNDP-FAO Regional Project RAS/86/022, pp. 10–14.

Ciapina LP and Lemos EGM (2000). Molecular characterization of the 9a5c isolate of *Xyllela fastidiosa* by rep-PCR. *Summa Phytopathologica* 26 (2): 262–264.

Civerolo EL (1981). *Citrus* bacterial canker disease: An overview. *Proc. Int. Soc. Citriculture,* vol. 1, pp. 390–394.

Civerolo EL (1984).Bacterial canker disease of *Citrus. J. Rio Grande Valley Hortic. Soc.* 37: 127–146.

Curry DW (1989). Eradication of citrus canker from Thursday Island. *Queensland Agric. J.* 115: 78–79.

De Lange JH (1978). Shoot tip grafting: A modified procedure. *Citrus and Subtropical Fruit Journal* 59: 13–15.

Filho Santos HP, de la Rosa Paguio O, da Cunha Sobrinho AP, da Silva Coelho Y, and Medina VM (1984). The *Citrus* variety improvement program in Brazil. *Proc. Int. Soc. Citriculture,* vol. 1, pp. 325–327.

Fu-ZhiHua, Xu-ZhanHua, Fu Zh and Xu ZH (2001). Study on the occurrence of *Citrus* canker and its control. *South China Fruits* 30 (4): 15.

Gabriel DW, Kingsley MT, Hunter JE, and Gottwald T (1989). Reinstatement of *Xanthomonas citri* (ex. Hasse) and *X. phaseoli* (ex. Smith) to species and reclassification of all *X. campestris* pv. *citri* strains. *Int. J. Syst. Bact.* 39: 14.

Gao S, Garnier M, and Bove, JM (1993). Production of monoclonal antibodies recognizing most Asian strains of greening BLO by in vitro immunization with an antigenic protein purified from the BLO. *Proc. 12th Conf. IOCV* (eds. P Moreno, JV da Graca, and LW Timmer). Univ. Calif. Riverside CA, pp. 245–249.

Garnier, M, and Bove JM (1993). *Citrus* greening disease and the greening bacterium. *Proc. 12th Conf. IOCV* (eds. P Romeno, JV da Gracia, and LW Timmer). Univ. Calif. Riverside CA, pp. 212–219.

Garnier M, and Bove JM (1997). Recent developments in vascular restricted walled bacteria of citrus: *Xyliela fastidiosa,* and the Libobacters, and Protobacterial plant pathogens. *Fruits.* 52 (6): 349–559.

Garnier M, Lalrille J, and Bove JM (1976). *Spiroplasma citri,* and the organism associated with Lukubin: Comparison of their envelope sytems. *Proc.7th Conf. IOCV* (ed. EC Calavan), Univ. Calif. Riverside CA, pp. 13–17.

Garnier M, Danel N, and Bove JM (1984). The greening organism is a Gram Negative Bacterium. *Proc. 9th Conf. IOCV* (eds. SM Garnsey, LW Timmer, and JA Dodds), Univ. Calif. Riverside CA, pp. 115–124.

Garnier M, Martin-Gros G, and Bove JM (1987). Monoclonal antibodies against the bacteria-like organism associated with *Citrus* greening disease. *Ann. Microbiol. Inst. Pasteur* 138: 639 - 650.

Garnier M, Chang CJ, Zreik L, Rossetti V, Bove JM (1993). *Citrus* variegated chlorosis: serological detection of *Xylellia fastidiosa,* the bacterium associated with the disease. *Proc. 12th Conf. IOCV* (eds. P Moreno, JV da Graca, and LW Timmer), Univ. Calif. Riverside CA, pp. 301–305.

Garnier M, Rao SJ, He YL, Villechanoux S, Gandar J, and Bove JM (1991). Study on Greening Organism (GO) with monoclonal antibodies: serological identification, morphology, serotype, and purification of the GO. *Proc. 11th Conf. IOCV* (eds. RH Brlansky, RF Lee, and LW Timmer). Univ. Calif. Riverside CA, pp. 428–435.

Goldschmidt EE (2000). Susceptibility of tangerines to citrus variegated chlorosis. *Proc. First Int. Symp. Citrus Biotechnology* (eds. WB Li, AJ Ayres, CX He, LC Donadio, R Goren). *Acta Horticulturae* 535: 253–257.

Goto M (1992). *Citrus* canker. *In:* Plant Diseases of International importance (eds. J Kumar, HS Chaube, US Singh, and AN Mukhopadhyay), Prentice Hall Int., London, UK, vol. III, pp. 170–208.

Goto M, Toyoshima A, and Messina MA (1980). A comparative study of the strains of *Xanthomonas campestris* cv. *citri,* isolated from citrus canker in Japan and cancrosis B in Argentina. *Ann. Phytopathol. Soc. Jpn.* 46: 329–338.

Gottwald TR, Graham JH, and Schubert TS (1997). An epidemiological analysis of the spread of citrus cankar in urban Miami, Florida, and synergistic interaction with the Asian *Citrus* leafminer. *Fruits* 52(6): 383–390.

Hartung JS, Daniel JF, and Provost O (1993). Detection of *Xanthomonas campestris* pv. *citri* by the polymerase chain reaction method. *Appl. Environ. Microbiol.* 59 (4): 1143–1148.

Hartung JS, Pruvost OP, Villemot I, and Atvarez A (1996). Rapid and sensitive colorimetric detection of *Xanthomonas axonopodis* pv. *citri* by immunocaptase and nested polymerase chain reaction assay. *Phytopathology* 86 (1): 95–101.

Hasse CH (1915). *Pseudomonas citri,* the cause of *Citrus* canker. *Jpn. Agric. Res* 4: 97–100.

He CX, Li WB, Ayres AJ, Hartung JS, Miranda VS, and Teixeira DC (2000). Distribution of *Xyllela fastidiosa* in *Citrus* rootstocks and transmission of *Citrus* variegated chlorosis between sweet orange plants through natural rootgrafts. *Plant Disease* 84 (6): 622–626.

Hung TingHsuan, Wu MengLing, Su HongJi, Hung TH, Wu ML and Su HJ (2001). Identification of Chinese bos (*Severinia boxifolia*) as an alternative host of the bacterium causing citrus Huan glung bin. *European J. Plant Path.* 107 (2): 183–109.

Igwegbe ECK and Calavan EC (1970). Occurrence of mycoplasma-like bodies in phloem of stubborn-infected *Citrus* seedlings. *Phytopathology* 60: 1525–1526.

Iwamoto M and Oku T (2000). Cloning and molecular characterization of *hrpx* from *Xanthomonas axonopodis* pv. *citri*. *DNA Sequence* 11 (1–2): 167–173.

Jagoueix S, Bove JM, and Garnier M (1994). The phloem-limited bacterium of greening disease of citrus is a member of the subdivision of the proteobacteria. *Inter. J. Syst. Bacreriol.* 44 (3): 386–397.

Ke C and Xua CF (1990). Successful integrated management of huan glung bin disease in several farms of Juangdung and Fujian by combining early eradication with integrated insecticides spraying. *In*: *Proc. 4th Int. Asia Pacific Conf. on Citrus Rehabilitation*: 145–148.

Lallemand JD, Fos, and Bove JM (1986). Transmission de la bacteria associee a la forme le psylle asiatique *Diaphorina citri* Kw. *Fruits*: 334–343.

Lama TK, Regmi C and Aubert B (1988). Distribution of citrus greening disease vector (*Diaphorina citri* Kuwayama) in Nepal, and to establishing biological control against it. *Proc. 10th Conf. IOCV* (eds. LW Timmer, SM Garnsey, and LV Navarro). Univ. Calif. Riverside CA, pp. 255–257.

Laranjeira FF, Pompeu Jr J, and Palazzo D (2000). Seeds from sweet orange fruit 'Natal' affected by citrus variegated chlorosis: germination, seedling growth and non-transmission of *Xyllela fastidiosa*. *Laranja* 21 (1): 161–173.

Li WB, Pria WD Jr, Teixeira DC, Miranda VS, Ayers AJ, Franco CF, Costa MG et al. (2001). Coffee leaf scorch caused by a strain of *Xyllela fastidiosa* from *Citrus*. *Plant Disease* 85 (5): 501–505.

Lopes JRS, Beretta MJG, Haracava R, Almeida RP, Krugner R, and Gracia Jr A (1996). Confirmacao da ansmissao por cigarrinhas do agente causal da clorose variegada dos citros, *Xylella fastidiosa*. *Fitopatol bras* 21 (suplemento): 343.

Luyi Luo and Yuan Chengdong (1991). Study on effective technique for eradicating *Citrus* canker. *In*: Rehabilitation of Citrus Industry in the Asia and Pacific Region (eds. Ke Chung, and SB Osman). UNDP-FAO Regional Project RAS/86/022: 196–199.

Masonie G, Garnier GM, and Bove JM (1976). Transmission of Indian *Citrus* decline by *Trioza erytreae* (Del Guercio), the vector of South African greening. *Proc. 7th Conf. IOCV* (ed. EC Calavan). Univ. Calif. Riverside CA, pp. 15–20.

Massina MA (1980a). Los metodos serlogicos en el estudo de la bacteria que produce la "cancrosis citrica" en la Argentina, III. Differenciation entre las bacterios de la "cancrosis A" o "asiatica", las de la "cancrosis B" o "sudamerica" en la pai. *INTA, Concordia, Estacion Experimental Agropecularis*, serie Tecnica 50: 3–8.

Massina MA (1980b). Los metodos serologicos en el estudo de la bacteria que produce la "cancrosis citrica" en la Argentina, IV. La tecnica de laboratorio. Resumenes II Cong. Nac. Citricultura, 23–31 October 1980, Concordia (E.R.) Argentina, p. 52.

Matsumoto T, Wang MC, and Su HJ (1961). Studies on likubin *Proc. 2nd Conf. IOCV* (ed. WC Price). Florida Univ. Press, Gainesville CA, pp. 121–125.

McClean APD and Oberholzer PCJ (1965). Citrus psyllid,a vector of the greening disease of sweet orange. *S. African J. Agr. Sci.* 8: 297–298.

Mehta A, Leite Jr RP and Rosato YB (2001). Assessment of the genetic diversity of the *Xyllela fastidiosa* isolated from citrus in Brazil by PCR-RFLP of the 16s rDNA and 16s-23s intergenic spacer and rep-PCR fingerprinting. *Antonie van Leeuwenhoek* 79 (1): 53–59.

Mitra R (1995). Studies of citrus canker (caused by *Xanthomonas campestris* pv. *citri*) and identification of sources of resistance to the bacterium. Ph.D. thesis, Genetics and Plant Breeding, Faculty of Agriculture, Bidhan Chandra Krishi Viswavidyalaya, West Bengal, India.

Monteiro PB, Renaudin J, Jagoueix-Eveillard S, Ayers AJ, Garnier M, and Bove JM (2001). *Catharanthus roseus*, an experimental host plant for the citrus strain of *Xyllela fastidiosa*. *Plant Disease* 85 (3): 246–251.

Montverde EE, Laborem G, Royes FJ, Ruiz R Jr, and Espinoza ME (1992). Research on citrus improvement in Venezuela. *Proc. Int. Soc. Citriculture*, vol. 2, pp. 752–755.

Mukhopadhyay S, Gurung A, Dutta Roy S, Sarkar TK, and Pradhan Jaishree (1992). Management of *Citrus* greening disease in Darjeeling hills. *In*: Management of Plant Diseases Caued by Fastidious Prokaryots (ed. SP Raychaudhuri). Assoc. Publ. Co., New Delhi, pp. 35–48.

Mukhopadhyay S, Rai (Pradhan) J, Sharma BC, Gurung A, Sengupta RK, and Nath PS (1996). Micropropagation of Darjeeling orange (*Citrus reticulata* Blanco) by shoot tip grafting. *J. Hort. Sci.*, 72 (3): 493–499.

Mukhopadhyay S, Thapa K, Mitra R, Thapa A, Rai J, and Gurung A (2000). Medicinal values of citrus, and integrated management of its diseases. Proc. Int. Conf. on Integrated Plant Disease Management for Sustainable Agriculture. *Indian Phytopath. Soc.* 2: 1225–1232.

Murray RGE and Schleifer KH (1994).Taxonomic note: a proposal for recording the properties of putative taxa of procaryots. *Int. J. Syst. Bacteriol.* 44: 174–176.

Namekata T and de Oliveira AR (1972). Comparative serological studies between *Xanthomonas citri* and a bacterium causing canker on Mexican lime: Proc. 3rd Int. Conf. Plant Pathog. Bact. Wageningen, The Netherlands, p. 365.

Navarro L (1981). *Citrus* shoot-tip grafting *in vitro* (STG) and its applications—A review. *Proc. Int. Soc. Citriculture* 1: 452–456.

Nariani TK, Raychaudhuri SP and Bhalla RB (1967). Greening virus of *Citrus* in India. *Indian Phytopath.* 20: 146–150.

Nicoli M (1984). La generation des agrumes en Corse par la technique de microgreffase de meristem es *vitro*. *Fruits* 40(2): 113–136.

Prommintara M (1988). Indexing greening in Thailand *Proc. 2nd FAO-UNDP Regional Workshop on Citrus Greening*, Lipa, Philippines, pp. 87–88.

Pruvost O, Hartung JS, Civerolo EL, Dubois C, and Perrier X (1992). Plasmid DNA fingerprints distinguish pathotypes of *Xanthomonas campestris* pv *citri*, the causal agent of *Citrus* bacterial canker disease. *Phytopathology* 82 (4): 485–490.

Pruvost O, Verniere G, Hartung J, Gottwald T, and Quetelard H (1997). Towards an improvement of *Citrus* cankar control in Reunion Island. *Fruits* 52 (6): 375–382.

Qin XT and Hartung JS (2001). Construction of a shuttle vector and transformation of *Xylella fastidiosa* with plasmid DNA. *Current Microbiol.* 43 (3): 158–162.

Qin XiaoTing, Miranda VS, Machado MA, Lemos EGM, Hartung JS, and Qin XT (2001). An evaluation of the genetic diversity of *Xyllela fastidiosa* isolated from diseased citrus and coffee in Sao Paulo, Brazil. *Phytopathology* 91 (6): 599–605.

Roberto SR, Coutinho A, De Lima JEO, Miranda VS, and Carlos EF (1996). Transmissao de *Xylella fastidiosa* pelas cigarrinhas *Dilobopterus costalimai, Acrogonia terminalis* e *Oncometopia facialis* em citros. *Fitopatol. Bras* 21 (4): 517.

Rodriguez GS, Garza-Lopez JG, Stapleton JJ, and Civerolo EL (1985). *Citrus* bacteriosis in Mexico. *Plant Dis.* 69: 808–810.

Rosetti V, Feichtenberger E, and Silveria ML (1982). Citrus Canker (*Xanthomonas campestris* pv *citri*): An analytical bibliography. Inst. Biol. Sao Paulo, Brazil.

Rosetti V, Garnier M, Bove JM, Beretta MJG, Teixeria ARR, Quaggio JA, and De Negri JD (1990). Presence de bacteres dans le xyleme d'orangers attaineints de chlorose variegee, une nouvelle maladie des agrumes au Bresil. *CR Acad. Sci. Paris* 310 (SerieIII): 345–349.

Saglio P, Lhospital M, Laflecha D, Dupont G, Bove JM, Tully JG, and Freundt EA (1973). *Spiroplasma citri*, gen and sp. n: a mycoplasma-like organism associated with "stubborn" disease of citrus. *Int. J. Syst. Bact.* 23: 191–204.

Salibe AA and Cortez RE (1968). Leaf mottling—a serious disease of *Citrus* in the Philippines. *Proc. 4th Conf. IOCV* (ed. JFL Childs). Univ. Florida Press, Gainesville CA, pp. 131–136.

Sanchez-Anguiano HM, and Felix-Astro FA (1984). An overview of *Citus* canker (bacteriosis) on Mexican lime at Tecoman, Colima, Mexico. *Proc. Int. Soc. Citriculture*, vol. 1, pp. 323–324.

Schoulties CL, Civerolo EL, Miller JW, Stall RE, Krass CJ, Poe SR, Ducharme, EP (1987). *Citrus* canker in Florida. *Plant Dis.* 71: 388.

Schwarz RE (1968). Thin layer chromatographical studies on phenolic markers of the greening virus in several citrus species. *South African J. Agric. Sci.* 11: 797–801.

Silva ER de, Vettore AL, Kemper EL, Leite A, Arrunda P and de Silva FR (2001). Fastidian gum: the *Xyllela fastidiosa* exopolysaccharide possibly involved in bacterial pathogenicity. *FEMS Microbiology News Letters* 203 (2): 165–171.

Simpson AJG, Reinach FC, Arruda P, Abreu FA, Acencio M, Alvarenga R et al. (2000). The genome sequence of the plant pathogen *Xyllela fastidiosa*. *Nature* 406: (6792): 151–157.

Singh B, Anitha K, Sharma KD, Agarwal PC, and Ram Nath (1995). World distribution of phytopathogenic bacteria. Division of Plant Quarantine, National Bureau of Plant Genetic Resources, New Delhi.

Sodhi SS, Dhillon SS, Jeyarajan R, and Chema SS (1973). Isolation and characterization of mycoplasma-like organisms associated with citrus greening disease. *Indian Jour. Microbiol.* 13: 13–16.

Stall RE and Civerolo EL (1991). Research related to the recent outbreak of *Citrus* canker in Florida. *Annu. Rev. Phytopathol.* 29: 399–420.

Stall RE, Miller JW, Marco GM, and Canteros de Echenique BI (1980). Population dynamics of *Xanthomonas citri* causing cancrosis of citrus in Argentina. *Proc. Fla. State Hortic. Soc.* 93: 10–14.

Sui Ke (1988). Attempt to obtain purified extract of greening (Huan glung bin) organism. *Proc. 2nd FAO-UNDP Regional Workshop*, Lipa, Philippines, pp. 90–94.

Tanaka S and Poi Y (1974). Studies on mycoplasma-like organism, suspected cause of citrus likubin and leaf-mottling. *Bull. Fac. of Agric.*, Tsmagwa Univ. 14: 64–70.

Tirtawidjaja S (1980). *Citrus* virus research in Indonesia. *Proc. 8th Conf. IOCV* (eds. EC Calavan, SM Garnsey, and LW Timmer). Univ. Calif. Riverside CA, pp. 129–132.

Varma A, Ahlawat YS, Chakraborty NK, Garnier M, and Bove JM (1993). Detection of greening BLO by electron microscopy, DNA hybridization in citrus leaves with or without mottle from various regions of India. *Proc. 12th Conf. IOCV* (eds. P Moreno, JV de Graca, and LW Timmer). Univ. Calif. Riverside CA, pp. 280–285.

Villechanoux S, Garnier M, Laigret F, Renaudin J, and Bove JM (1992). The genome of the non-cultured, bacteria-like organism associated with *Citrus* greening disease contained in the nusG-rplKAJL-ropBC gene cluster and the gene for a bacteriophage type DNA polymerase. *Curr. Microbiol.* 26: 161–166.

Weerawut P, Intrawimolsri S, and Boonyong C (1987). Survey of citrus aphids. *Citrus* psyllid, and citrus leaf-miner. *In:* Research Report of Entomology and Zoology Division, Department of Agriculture (in Thai).

Wells JM, Raju BC, Hung HY, Weisberg WG, Man-delco-Paul L, and Brenner DJ (1987). *Xylella fastidiosa* gen nov, sp nov: Gram negative, xylem-limited, fastidious plant bacterium related to *Xanthomonas* spp. *Int. J. Syst. Bacteriol.* 37: 136–143.

Winarno, M and Supriyanto A (1991). Disease-free material propagation, and nursery management. *In:* 6th Asia Pacific UNDP-FAO Conf. on Integrated Citrus Health Management (eds. C Ke and MO Shamshudin), pp. 21–25.

Wu Shih-Pan (1987). A direct fluorescence detection method for the diagnosis of *Citrus* yellow shoot. *In: Proc. Regional Workshop on Citrus Greening Huan glung bin,* Fuzhou, China.

Wu Shih-Pan, and Hwei-Chung Faan (1987). A microscopic method for rapid diagnosis of *Citrus* yellow shoot disease. *In: Proc. Regional Workshop on Citrus Greening Huan glung bin,* Fuzhou, China.

Wu Shih-Pan and Hwei-Chung Faan (1988). Histochemical diagnosis of citrus yellow shoot. *Jour. South China Agric. Univ.* 9: 79–80.

Xu CF, Xia WH, Li KB, and Ke C (1988). Further study of the transmission of *Citrus* Huan glung bin by a psyllid, *Diaphorina citri* Kuwayama. *Proc. 10th Conf. IOCV* (eds. LW Timmer, SM Garnsey, and L Navarro). Univ. Calif. Riverside CA, pp. 243–248.

C. Spiroplasmal Diseases

Bove JM (1980). Characterization of Spiroplasmas by polyacrylamide gel analysis of their proteins and enzymes. *Proc. 8th Conf. IOCV* (eds. EC Calavan, SM Garnsey, LW Timmer), Univ. Calif. Riverside CA, pp. 133–144.

Bove JM (1988). *Spiroplasma citri*: fifteen years of research. *Proc. 10th Conf. IOCV* (eds. LW Timmer, SM Garnsey, L Navarro), Univ. Calif. Riverside CA, pp. 274–284.

Bove JM and Garnier M (1997). Recent developments in phloem-restricted wall-less bacteria of *Citrus*: *Candidatus Phytoplasma aurantifolia* and *Spiroplasma citri*, two mycoplasmal plant pathogens. *Fruits* 52 (6): 349–359.

Bove JM, Garnier M, Mjeni AM, and Khayrallah A (1988). Witches' broom disease of small-fruited acid lime trees in Oman: first MLO disease of *Citrus*. *Proc. 10th Cnf. IOCV* (eds. LW Timmer, SM Garnsey, and l Navarro), Univ. Calif. Riverside A, pp. 307–309.

Calavan EC and Oldfield GN (1979). Symptomatology of spiroplasmal plant diseases. *In:* The Mycoplasmas (eds. RF Whitecomb and JG Tully), Academic Press, New York, NY, vol. III, pp. 37–64.

Cole RM, Tully JG, Popin TJ, and Bove JM (1973). Morphology, ultrastructure and bacteriophage infection of the helical mycoplasmas (*Spiroplasma citri* gen. nov. sp. nov.) cultured from "stubborn" disease of *Citrus*. *J.. Bact.* 115: 367–386.

Markham PG and Townsend R (1974).Transmision of *Spiroplasma citri* to plants. *Colloq. Inst. Natl. Sante Rech. Med.* 33: 201–206.

Mouches C, Duthil P, Vignault JC, Protopapadakis F, Nhmi A, Tully JG, and Oldfield GN, Sullivan DA, and Calavan EC (1984). Aspects of transmission of the *Citrus* stubborn pathogen by *Scaphytopius nitridus* (DeLong). *Proc. 10th Conf. IOCV* (eds. LW Timmer, SM Garnsey, and L Navarro), Univ. Calif. Riverside CA, pp. 131–136.

Oldfield GN, Kaloostian GH, Pierce HD, Granett AL and Blue RL (1976). Beet leafhoppers transmit *Citrus* stubborn disease. *Calif. Agr.* 30 (6): 15.

Olson EO and Rogers B (1969). Effects of temperature on expression and transmission of stubborn disease of *Citrus*. *Plant Disease Rep.* 53: 45–49.

Roistacher CN and Calavan EC (1972). Heat tolerance of preconditioned *Citrus* budwood for virus inactivation. *Proc. 5th Conf. IOCV* (ed. WC.Price), Univ. Florida Press, Gainesville, FL. pp. 256–261.

Sagilo P, Lhospital M, Lafleche D, Dupont G, Bove JM, Tully JG and Freundt EA (1973). *Spiroplasma citri* gen. sp: a mycoplasma-like organism associated with "stubborn" disease of *Citrus*. *Int. J. Syst. Bact.* 23(3): 191–204.

Saillard CO, Garcia-Jurado O, Bove JM, Vignault JC, Moutous TG, Fos A, Bonfils A et al. (1980). Application of ELISA to the detection of *Spiroplasma citri* in plants and insects. *Proc. 8th Conf. IOCV* (eds. EC.Calavan, SM Garnsey, and LW Timmer), Univ. Calif. Riverside CA, pp. 145–152.

Phytoplasmal Diseases

Bove JM ad Garnier M (2000). Witches' broom diseases of lime. *Arab Journal of Plant Protection* 18 (2): 148–152.

Viral and Viroid Diseases

Ahlawat YS, Pant RP, Lockhart BEL, Srivastava M. Chakraborty NK, Varma A (1996). Association of a Badna virus with *Citrus* mosaic disease in India. *Plant Disease* 80 (5): 590–592.
Albiach-Marti MR, Guerri J, Cambra M, Garnsey SM, amd Moreno P (2000). Differentiation of citrus tristeza virus isolates by serological analysis of p25 coat protein peptide maps. *J. Virological Methods* 88 (1): 25–34.
Albiach-Marti MR, Mawassi M, Gowda S, Satyanarayana T, Hilf ME, Shanker S et al. (2000a). Sequences of *Citrus* tristeza virus separated in time and space are essentially identical. *J. Virology* 74 (15): 6856–6865.
Albiach-Marti MR, Guerri J, Mendoza AH de, Laigret F, Bellester-Olmos JF, Moreno P, et al. (2000b). Aphid transmission alters the genomic and defective RNA populations of *Citrus* tristeza virus isolate. *Phytopathology* 90 (2): 134–138.
Baksh N, Lee RF, and Garnsey SM (1984). Detection of *Citrus* exocortis viroid by plyacrylamide gel electrophoresis. *Proc. 9th Conf. IOCV* (eds. SM Garnsey, LW Timmer, and JA Dodds). Univ. Calif., Riverside, CA, pp. 343–352.
Balaraman K (1980). Interaction studies between a mild strain of tristeza on acid lime and other viruses, and virus-like diseases of *Citrus. Proc. 8th Conf. IOCV* (eds. EC Calavan, SM Garnsey, and LW Timmer). Univ. Calif., Riverside, CA, pp. 54–59.
Balaraman K and Ramakrishnan K (1980). Variation and cross protection in *Citrus* tristeza virus on acid lime. *Proc. 8th Conf. IOCV* (eds. EC Calavan, SM Garnsey, and LW Timmer). Univ. Calif., Riverside, CA, pp.60–80.
Bar-Joseph M and Nitzan Y (1991). The spread and distribution of *Citrus* tristeza virus isolates in sour orange seedlings. *Proc. 11th Conf. IOCV* (eds. RH Brlansky, RF Lee, and LW Timmer). Univ. Calif., Riverside, CA, pp.162–165.
Bar-Loseph M, Loebenstein G, Oren Y (1974). Use of electronmicroscopy in eradication of tristeza sources recently found in Israel. *Proc. 6th Conf. IOCV* (eds. LG Weathers and M Cohen) Univ. Calif. Div. Agric. Sciences, Berkeley, CA, pp. 83–85.
Bar-Joseph M, Garnsey SM, Gosalves D, and Purcifull DE (1980). Detection of *Citrus* tristeza virus. I: Enzyme-linked immunosorbent assay (ELISA) and SDS-immunodiffusion methods. *Proc. 8th Conf. IOCV* (eds. EC Calavan, SM Garnsey and LW Timmer). Univ. Calif., Riverside, CA, pp. 1–8.
Boccardo G, La Rosa R, and Catara A (1984). Detection of *Citrus* exocortis viroid by polyacrylamide gel electrophoresis of nucleic acid from greenhouse citrus. *Proc 9th Conf. IOCV* (eds.P Moreno, JV daGraca, and LW Timmer). Univ. Calif., Riverside, CA, pp. 357–361.
Bove C, Vogel R, Albertini D, and Bove JM (1988).Discovery of a strain of tristeza virus (K) inducing no symptom in Mexican lime. *Proc. 10th Conf. IOCV* (eds. LW Timmer, S M Garnsey, and L Navarro). Univ. Calif., Riverside, CA, pp. 17–21.
Brlansky RH, Pelosi RR, Garnsey SM, Youtsey CO, Lee RF, Yokomi RK, and Sonoda RM (1986). Tristeza quick decline epidemic in south Florida. *Proc. Fla. State Hort. Soc.* 99: 66–69.
Byadgi AS, Ahlawat YS, Chakraborty NK, Varma A, Srivastava M, and Milne RG (1993). Observation of a filamentous virus associated with *Citrus* ringspot in India. *Proc. 12th Conf. IOCV* (eds. P Moreno, JV da Graca, and LW Timmer). Univ. Calif., Riverside, CA, pp. 155–162.

Calavan EC, Harjung MK, Blue RL, Roistacher CN, Gumpf DJ, and Moore PW (1980). Natural spread of Seedling Yellows and sweet orange and grapefruit Stempitting tristeza viruses at the University of California, Riverside. *Proc.8th Conf. IOCV* (eds. EC Calavan, SM Garnsey, and LW Timmer). Univ. Calif., Riverside, CA, pp. 69–75.

Calvert LA, Lee RF, and Hiebert E (1988). Characterization of the RNA species of *Citrus* variegation virus with complementary DNA clones. *Proc.10th Conf. IOCV* (eds. LW Timmer, SM Garnsey, L Navarro). Univ. Calif., Riverside, CA, pp. 327–333.

Cambra M, Camarasa E, Gorris MT, Garnsey SM, and Carbonell E (1991). Comparison of different immunosorbent assays for *Citrus* tristeza virus (CTV) using CTV-specific monoclonal and polyclonal antibodies. *Proc. 11th Conf. IOCV* (eds. Brlansky RH, Lee RF, and Timmer LW). Univ. Calif., Riverside, CA, pp. 38–45.

Cambra M, Camarasa E, Gorris MT, Garnsey SM, Gumpf DJ, and Tsai MC (1993). Epitope diversity of *Citrus* tristeza virus isolates in Spain. *Proc.12th Conf. IOCV* (eds. P Moreno, JV da Graca, LW Timmer). Univ. Calif., Riverside, CA, pp. 33–38.

Chagas CM, Rossetti Victoria, and Chiavegato LG (1984). Transmission of leprosies by grafting. *Proc. 9th Conf. IOCV* (eds. SM Garnsey, LW Timmer, and JA Dodds). Univ. Calif., Riverside, CA, pp. 215–217.

Chakraborty NK, Ahlawat YS, Varma A, Jagadish Chandra K, Rampandu S, and Kapur P (1993). Serological reactivity of *Citrus* tristeza virus strains in India. *Proc.12th Conf. IOCV* (eds. P Moreno, JV daGraca, and LW Timmer). Univ. Calif., Riverside, CA, pp. 108–112.

Che-Xi Bing, Piestun D, Mawassi M, Young-GuAng, Satyanarayana T, Gowda S, et al. (2001). 5′-co-terminal sub-genomic RNAs in *Citrus* tristeza virus-infected cells. *Virology* 283 (2): 374–385.

Chung Ke and Ru-Jian Wu (1991). Occurrence and distribution of *Citrus* tatterleaf in Fujian, China. *Proc. 11th Conf. IOCV* (eds. RH Brlansky, RF Lee, and LW Timmer). Univ. Calif., Riverside, CA, pp. 358–364.

Corbett MK and Price WC (1967). Failure of psorosis to cross protect against *Citrus* variegation virus. *Fla. Agri. Exp. J.* ser No 2415, 83: 151–153.

D'Onghia AM, Djelouach K and Savino V (2000). Serological detection of *Citrus* psorosis virus in seeds but not in seedlings of infected mandarin and sour orange. *J. Plant Pathology* 83 (3): 233–235.

D'Onghia AM, Djelouach K, Frasheri D and Potere O (2001). Detection of *Citrus* psorosis virus by direct tissue blot immunoassay. *J. Plant Pathology* 83(2): 139–142.

d'Uorso F, Ayllon MA, Rubio L, Sambade A, Mendoza AH de, Guerri J, Moreno P, and deMendoza AH (2000). Contribution of uneven distribution of genomic RNA variants of *Citrus* tristeza virus (CTV) within the plant to changes in the viral population following aphid transmission. *Plant Pathology* 49 (2): 288–294.

DaGraca JV and Maharaj SB (1991). *Citrus* vein enation virus—a probable luteovirus. *Proc.11th Conf. IOCV* (eds. RH Brlansky, RF Lee, and LW Timmer). Univ. Calif., Riverside, CA, pp. 391–394.

DaGraca JV, Marais LJ, and Broembsen LA (1982). Severe Stempitting in young grapefruit. *Citrus and Subtrop. Fruit J.* 588: 18–21.

DaGraca JV, Marais LJ, and Von Broembsun A (1984). Severe tristeza stempitting decline of young grapefruit in South Africa. *Proc 9th Conf. IOCV* (eds. SM Garnsey, LW Timmer, and JA Dodds). Univ. Calif., Riverside, CA, pp. 62–65.

DaGraca JV, Bar-Joseph M, and Derrick KS (1993). Immunoblot detection of *Citrus* psorosis using *Citrus* ringspot antiserum. *Proc.12th Conf. IOCV* (eds. P Moreno, JV da Graca, and LW Timmer). Univ. Calif., Riverside, CA, pp. 432–434.

DaGraca JV, Lee RF, Moreno E, Civerolo L, and Derrick KS (1991). Comparison of isolates of *Citrus* ringspot, psorosis, and other virus-like agents of *Citrus*. *Plant Dis.* 75: 613–616.

Davino M and Garnsey SM (1984). Purification, characterization, and serology of a mild strain of *Citrus* infectious variegation virus from Florida. *Proc. 9th Conf. IOCV* (eds. SM Garnsey, LW Timmer, and DA Dodds). Univ. Calif., Riverside, CA, pp. 196–203.

Derrick KS, Barthe GA, Hewitt BG, and Lee RF (1993). Serological tests for *Citrus* blight. *Proc.12th Conf. IOCV* (eds. P Moreno, JV da Graca, and LW Timmer). Univ. Calif., Riverside, CA, pp. 121–126.

Derrick KS, Brlansky RH, Lee RF, Timmer LW, and Nguyen TK (1988). Two components associated with *Citrus* ringspot virus. *Proc.10th Conf. IOCV* (eds. LW Timmer, SM Garsney, and L Navarro). Univ. Calif., Riverside, CA, pp. 340–342.

Diener TO (1979). Viroid and Viroid Diseases. Wiley Interscience, New York, NY.

Duran-Vila N, Pina JA, and Navarro L (1993). Improved indexing of *Citrus* viroids. *Proc.12th Conf. IOCV* (eds. SM Garnsey, LW Timmer, and JA Dodds). Univ. Calif., Riverside, CA, pp. 202–211.

Duran-Vila N, Pina JA, Ballester JF, Juarez J, Roistacher CN, Rivera-Bustamante R, and Semancik JS (1988). The *Citrus* exocortis disease: a complex of viroid RNAs. *Proc.10th Conf. IOCV* (eds. LW Timmer, SM Garnsey, and L Navarro). Univ. Calif., Riverside, pp. 152–164.

Fawcett HS (1911). Scaly bark or nail-head rust of *Citrus*. *Florida Agr. Expt. Sta. Bull.* 106: 3–41.

Fawcett HS (1933). New symptoms of psoroses indicating a virus disease of *Citrus*. *Phytopathology* 23: 93 (Abstr).

Foissac X and Duran-Vila N (2000). Characterization of a *Citrus* apscaviroids isolated in Spain. *Archives of Virology* 145 (9): 1975–1983.

Fraser LR, Levitt EC, and Cox J (1961). Relationship between exocortis and stunting of citrus varieties on *Poncirus trifoliata* rootstock. *Proc. 2nd Conf. IOCV* (ed. WC Price). Univ. Florida Press, Gainesville, Fla, pp. 34–39.

Frolich EF, Calavan EC, Carpenter JB, Christiansen DW, and Roistacher CN (1965). Differences in response of citron selections to exocortis virus infection. *Proc. 3rd Conf. IOCV* (ed. WC Price). Univ. Florida Press, Gainesville, Fla, pp. 113–118.

Galipienso L, Vives MC, Moreno P, Milne RG, Navarro L, and Guerri J (2001). Partial characterization of *Citrus* leaf blotch virus, a new virus from Nagami kumquat. *Archives of Virology* 146 (2): 357–368.

Garnsey SM and Timmer LW (1980). Mechanical transmissibility of *Citrus* ringspot virus isolates from Florida, Texas, and California. *Proc. 8th Conf. IOCV* (eds. SM Garnsey and LW Timmer). Univ. Calif., Riverside, CA, pp. 174–179.

Garnsey SM and Timmer IW (1988). Local lesin isolate of *Citrus* ringspot virus induces psorosis bark scaling. *Proc.10th Conf. IOCV* (eds. LW Timmer, SM Garnsey, and L Navarro). Univ. Calif., Riverside, CA, pp. 334–339.

Garnsey SM, Gonsalves D, and Purcifull DE (1978). Rapid diagnosis of *Citrus* tristeza virus infections by sodium dodecyl sulphate-immunodiffusion procedures. *Phytopathology* 68: 88–95.

Garnsey SM, Christie RG, Derrick KS, and Bar-Joseph M (1980). Detection of *Citrus* tristeza virus. II: Light and electron microscopy of inclusions and viral particles. *Proc. 8th Conf. IOCV* (eds. EC Calavan, SM Garnsey, and LW Timmer). Univ. Calif., Riverside, CA, pp. 9–16.

Garnsey SM, Permar TA, Cambra M, and Hendeson CT (1993). Direct tissue blot immunoassay (DTBIA) for detection of *Citrus* tristeza virus (CTV). *Proc.12th Conf. IOCV* (eds. P Moreno, JV da Graca, and LW Timmer). Univ. Calif., Riverside, CA, pp. 39–50.

Garnsey SM, Permar TA, Cambra M, Koizumi M, and Vela C (1989). Epitope diversity among *Citrus* tristeza virus isolates. *Phytopathology* 79: 1174 (Abstr.).

Goldschmidt EE (2000). Molecular marker linked to *Citrus* tristeza virus (CTV) resistance gene in cultivars of *Poncirus trifoliata*. Proc.1st Int. Symp. Citrus Biotechnology (eds. M Cristofani, Machade MA and Goren R). *Acta Horticulturae* 535: 231–235.

Gonsalves D and Garnsey SM (1975). Nucleic acid components of *Citrus* variegation virus and their activation by coat protein. *Virology* 67: 311–318.

Gracia ML, Arrese EL, Grau O, and Sarachu AN (1991). *Citrus* psorosis disease agent behaves as a two-component-ssRNA virus. *Proc. 11th Conf. IOCV* (eds. RH Brlansky, RF Lee, and LW Timmer). Univ. Calif., Riverside, CA, pp. 337–344.

Grasso S (1973). Simultaneous infection by ringspot, infectious variegation, and psorosis A viruses on lemon plants. *Technica Agricola*, Italy 25: 329–336.

Guerri JP, Moreno R, and Lee RH (1990). Identification of *Citrus* tristeza virus strains by peptide maps of virus coat protein. *Phytopathology* 80: 682–698.

Gurung Anita (1989). Ecology of vectors of *Citrus* tristeza virus and greening in Darjeeling district and its relations with the spread of citrus dieback. Ph.D. thesis, Bidhan Chandra Krishi Vishwavidyalaya, West Bengal, India.

Hailstones DL, Bryant KL, Broadbent P and Zhou C (2000). Detection of tatterleaf virus with reverse transcription-polymerase chain reaction (RT-PCR). *Australian Plant Pathology* 29 (4): 240–248.

Hermoso de Mendoza, Bellester-Olmos JF, and Pina JA (1988). Spread of *Citrus* tristeza virus in a heavily infested citrus area in Spain. *Proc.10th Conf. IOCV* (eds. LW Timmer, SM Garnsey, and L Navarro). Univ. Calif. Riverside, CA, pp. 68–70.

Hermoso de Mendoza, Bellester-Olmos JF, and Pina JA (1988a). Differences in transmission efficiency of *Citrus* tristeza virus by *Aphis gossypii* using sweet orange, mandarin, and lemon trees as donor or receptor host plants. *Proc.10th Conf. IOCV* (eds. LW Timmer, SM Garnsey, and L Navarro). Univ. Calif., Riverside, CA, pp. 62–64.

Hermoso de Mendoza A, Pina JA, Bellester-Olmos JF, and Navarro L (1993). Persistent transmission of *Citrus* vein enation virus by *Aphis gossypii*, and *Myzus persicae*. *Proc.12th Conf. IOCV* (eds. P. Moreno, JV da Graca, and LW Timmer). Univ. Calif., Riverside, CA, pp. 361–363.

Hilf ME, Karasev AV, Albiach-Marti MR, Dowson WO, and Garnsey SM (1999). Two paths of sequence divergence in *Citrus* tristeza virus complex. *Phytopathology* 89(4): 336–342.

Hiroyuki Ieki and Yamada Shun-ichi (1980). Inactivation of *Citrus* tristeza virus (CTV) with heat treatment: Heat tolerance and inactivation of CTV on root-scion combinations. *Proc. 8th Conf. IOCV* (eds. EC Calavan, SM Garnsey, and LW Timmer). Univ. Calif., Riverside, CA, pp. 20–24.

Ito T, Teki H, and Ozaki T (2001). Characteriziation of a new *Citrus* viroid species tentatively termed citrus viroid OS. *Archives Virology* 146 (5): 975–982.

Iwanami T (2000). Taxonomic studies of Satsuma dwarf virus and related viruses. *Bull. Nat. Inst. Fruit Sci.* 34: 153–157.

Iwanami T, Omura M, and Ieki H (1993). Susceptibility of several *Citrus* relatives to Satsuma dwarf virus. *Proc.12th Conf. IOCV* (eds. P Moreno, JV da Graca, and LW Timmer). Univ. Calif., Riverside, CA, pp. 352–356.

Iwanami T, Kano T, Koizumi M, and Watanabe Y (1991). Leaf blotch and stem mosaic symptoms in sweet orange (*Citrus sinensis* Osbeck) caused by astrain of Satsuma dwarf virus (SDV). *Bull. Fruit Trees Res. Stn.* 21: 75–83.

Iwanami T, Kondo Y, Kobayashi T, Han SS and Karasev AV (2001). Sequence diversity and interrelationships among isolates for Satsuma dwarf related viruses. *Archives of Virology* 146 (4): 807–813.

Jacomino Angelo P and Salibe AA (1993).Occurrence of woody gall disease in *Citrus* in Sao Pauolo State, Brazil. *Proc. 12th Conf. IOCV* (eds.) P Moreno, JV daGraca, and LW Timmer). Univ. Calif., Riverside, CA, pp. 357–360.

Jarupat T, Dodds JA, and Roistacher CN (1988). Effect of host passage on ds RNAs of two strains of *Citrus* tristeza virus. *Proc.10th Conf. IOCV* (eds. LW Timmer, SM Garnsey, and L Navarro). Univ. Calif., Riverside, CA, pp. 39–45.

Kano T and Koizumi M (1991). Separation of *Citrus* tristeza virus (CTV) serotypes through aphid transmission. *Proc.11th Conf. IOCV* (eds. RH Brlansky, RF Lee, and LW Timmer). Univ. Calif., Riverside, CA, pp. 82–85.

Kano T, Garnsey SM, Koizumi M, and Permar TM (1991). Serological diversity of field sources of *Citrus* tristeza virus (CTV) in Japan. *Proc.11th Conf. IOCV* (eds. RH Brlansky, RF Lee, and LW Timmer). Univ. Calif., Riverside, CA, pp. 51–55.

Karasev AV, Han SS, and Iwanami T (2001). Satsuma dwarf and related viruses belong to a new lineage of plant picorna-like virus. *Virus Gene* 23 (1): 45–52.

Kitajima EW, Silva DM, Oliveira AR, Muller GW, and Costa AS (1963). Thread-like particles associated with tristeza of *Citrus*. *Nature* 201: 1011–1012.

Koizumi M and Kuhara S (1984). Protection of preinoculated *Citrus* trees against tristeza virus in relation to the virus concentration detection by ELISA. *Proc. 9th Conf. IOCV* (eds. SM Garnsy, LW Timmer, and JA Dodds). Univ. Calif., Riverside, CA, pp. 41–48.

Koizumi M and Kano T (1991). Improvement of *Citrus* nursery systems used for indexing of viruses. *Bull. Fruit Tree Res. Stn.* 20: 79–91.

Koizumi M, Kano T, Ieki H, and Mae H (1988). China laurestine: a symptomless carrier of Satsuma dwarf virus that accelerates natural transmission in the fields. *Proc.10th Conf. IOCV* (eds. LW Timmer, SM Garnsey, and L Navarro). Univ. Calif., Riverside, CA, pp. 348–352.

Kong Ping, Rubio L, Polek M, Falk BW and Kong P (2000). Population structure and genetic diversity within California *Citrus* tristez virus (CTV) isolates. *Virus Genes* 21 (3): 139–145.

Lee RF, Calvert LA, and Hubbard JD (1988). *Citrus* tristeza virus: characteristics of coat proteins. *Phytopathology* 78: 1221–1226.

Lee RF, Timmer LW, Purcifull DE, and Garnsey SM (1981). Comparison of radioimmunosorbent assays for detection of rickettsia-like bacteria and *Citrus* tristeza virus. *Phytopathology* 71: 889 (Abstr.).

Levy L and Gumpf DJ (1991). Studies on the psorosis disease of *Citrus*, and preliminary characterization of a flexuous virus associated with the disease. *Proc.11th Conf. IOCV* (eds. RH Brlansky, RF Lee, and LW Timmer). Univ. Calif., Riverside, CA, pp. 319–336.

Levy L and Hadidi A (1993). Direct nucleotide sequencing of PCR-amplified DNAs of the closely related *Citrus* viroids IIa, and IIb (cachexia). *Proc.12th Conf. IOCV* (eds. P Moreno, JV daGraca, and LW Timmer). Univ. Calif., Riverside, CA, pp, 180–186.

Lin YouJian, Rundell PA, Xie Lianhui, Powell CA, Lin YJ, and Xie LH (2000). In situ immunoassay for detection of *Citrus* tristeza virus. *Plant Disease* 84 (9): 937–940.

Lin YouJian, Xie Lianhui, Rundell PA, Powell CA, Lin YJ, and Xie LH (2000a). Development of western blot procedure for using polyclonal antibodies to study the proteins of citrus tristeza virus. *Acta Phytopathologic Sinica* 30 (3): 250–256.

Lovisolo O (2001). *Citrus* leprosies virus: properties, diagnosis, agro-ecology and phytosanitary importance. *Bull. OEPP* 31 (1): 79–89.

Maharaj SB and deGraca JV (1988). Observation of isometric virus-like particles associated with *Citrus* vein enation-infected *Citrus* and the viruliferous aphid vector *Toxoptera citricidus*. *Phytophylactica* 20: 357–360.

Maharaj SB and daGraca JV (1989). Transmission of *Citrus* vein enation virus by *Toxoptera citricidus*. *Phytopathologica* 21: 81–82.

Marco GM, Semancik JS, and Gumpf DJ (1991). Cucumo-like virus isolated from cowpea indicator plants manifesting the *Citrus* tatter leaf virus syndrome. *Proc.11th Conf. IOCV* (eds. RH Brlansky, RF Lee, and LW Timmer). Univ. Calif. Riverside, CA, pp. 352–357.

Martelli GP (1991). Immunosorbent electron microscopy. *In*: Graft-transmissible Diseases of Citrus (ed. CN Roistacher). IOCA, FAO, Rome, pp. 249–251.

McClean APD (1974).The tristeza complex. *Proc. 6th Conf. IOCV* (eds. LG Weathers and M Cohen) Univ. Calif., Riverside, CA, pp. 59–66.

Mei-Chen Tsai and Hong-ji Su (1991). Development and characterization of monoclonal antibodies to *Citrus* tristeza virus (CTV) strains in Taiwan. *Proc. 11th Conf. IOCV* (eds. RH Brlansky, RF Lee, and LW Timmer). Univ. Calif., Riverside, CA, pp. 46–50.

Milne RG (1988). The Plant Viruses. Plenum Press, New York, NY, vol. 4.

Miyakawa T and Tsuji M (1976). A bud-union abnormality of Satsuma mandarin on *Poncirus trifoliata* rootstock in Japan. *Proc. 7th Conf. IOCV* (ed. EC Calavan). Univ. Calif., Riverside, CA, pp. 125–131.

Miyakawa Tand Yamaguchi (1981). Tristeza stempitting in *Citrus* diseases in Japan. Japan Plant Protection Association, Komagome, Toshima, Tokyo, 117 pp.

Moreno P, Guerri J, and Munoz M (1990). Identification of Spanish strains of *Citrus* tristeza virus by analysis of double-stranded RNA. *Phytopathology* 80: 477–482

Narvaez G, Slimane-Skander B, Ayllon MA, Rubo L, Guerri J, and Moreno P (2000). A new procedure to differentiate *Citrus* tristeza virus isolates by hybridization with dioxygenin-labelled cDNA probes. *J. Virological Methods* 85 (1–2): 83–92

Navas-Castillo J and Moreno P (1993). *Citrus* Ringspot Diseases in Spain. *Proc. 12th Conf. IOCV* (eds. P Moreno, JV daGraca, and LW Timmer). Univ. Calif and Riverside, CA, pp. 163–172.

Navas-Castillo J, Moreno P, Bellester-Olmos JF, Pina JA, and Hermoso de Mendoza A (1991). Detection of a necrotic strain of *Citrus* ringspot in Star Ruby grapefruit in Spain. *Proc.11th Conf. IOCV* (eds. RH Brlansky, RF Lee, and LW Timmer). Univ. Calif., Riverside, CA, pp. 345–351.

Nishio T, Kawai A, Takahshi T, Namba S, and Yamashita S (1989).Purification and properties of *Citrus* tatterleaf virus. *Ann. Phytopath. Soc. Jpn.* 55: 254–258

Nolasco G, Sequeira Z, Bonacalza B, Mendes C, Torres V, Sanchez F, Urgoiti B, et al. (1997). Sensitive *Citrus* tristeza virus diagnosis using immuno-puncture, reverse transcriptional polymerase chain reaction and exonuclease fluorescence. *Fruits* 52 (6): 391–396.

Owens RA, Yang G, Gundersen-Rindal D, Hammond RW, Candrese T, and Bar-Joseph M (2000). Both point mutation and RNA recombination contribute to the sequence diversity of *Citrus* viroid III. *Virus Genes* 20 (3): 243–252.

Palacio Bielsa A, Foissac X, and Duran-Vila N (1999). Indexing of *Citrus* viroids by imprint hybridization. *Europ. J. Phytopathology* 105 (9): 897–903.

Permar TA and Garnsey SM (1988). A monoclonal antibody that discriminates strains of tristeza virus. *Phytopathology* 80: 224–228.

Pires Leitao TM, Romero J, and Duran-Vila N (1993). Detection of pathogenesis related proteins associated with viroid infection in *Citrus. Proc.12th Conf. IOCV* (eds. P Moreno, JV daGraca, and LW Timmer). Univ. Calif., Riverside, CA, pp. 106–201.

Polizzi G, Albanese G, Azzaro A, Davino N, and Catara A (1991). Field evaluation of dwarfing effect of two combinations of *Citrus* viroids. *Proc.11th Conf. IOCV* (eds. RH Brlansky, RF Lee, and LW Timmer). Univ. Calif., Riverside, CA, pp. 230–233.

Potere O, Boscia D, Djelouah K, Elicio V, and Savino V (1999). Use of monoclonal antibodies to *Citrus* psorosis virus for diagnosis. *J. Plant Pathology* 81 (3): 209–212.

Raccah B, Bar-Joseph M, Lobenstein G (1977). The role of aphid vectors and variation in virus isolates in the epidemiology of tristeza disease. Contribution from Volcani Center, Agr. Res. Organization, 273-E.

Rogers GM, daGraca JV (1986). Virus-like particles of citrus vein enation virus-infected tissue. *Proc. EM Soc. Southern Afr.* 16: 127–128.

Roistacher CN (1982). A Blueprint for disaster. Part 2: Changes in transmissibility of seedling yellows. *Citrograph* 67: 28–32.

Roistacher CN (1988). *Citrus* tatterleaf virus: further evidence for a single virus complex. *Proc.10th Conf. IOCV* (eds. LW Timmer, SM Garnsey, and L Navarro). Univ. Calif., Riverside, CA, pp. 353–359.

Roistacher CN (1993). Psorosis—A Review. *Proc.12th Conf. IOCV* (eds. P Moreno, JV daGraca, and LW Timmer). Univ. Calif., Riverside, CA, pp. 116–124.

Roistacher CN and Moreno P (1991). The worldwide threat from destructive isolates of *Citrus* tristeza virus—A review. *Proc. 11th Conf. IOCV* (eds. RH Brlansky, RF Lee, and LW Timmer). Univ. Calif., Riverside, CA, pp. 7–19.

Roistacher CN, Nauer EM, and Wagner RC (1980). Transmissibility of cachexia, Dweet mottle, psorosis, tatterleaf, and infectious variegation viruses on knife blades and its prevention. *Proc. 8th Conf. IOCV* (eds. EC Calavan, SM Garnsey, and LW Timmer). Univ. Calif., Riverside, CA, pp. 225–229.

Roistacher CN, Dodds JA, and Bash JA (1988). Cross protection against *Citrus* tristeza seedling yellows and stempitting viruses by protective isolates developed in greenhouse plants. *Proc.10thConf.IOCV* (eds. LW Timmer, SM Garnsey, and L Navarro). Univ. Calif., Riverside, CA, pp. 90–100.

Roistacher CN, Pehrson JE, and Semancik JS (1991). Effect of *Citrus* viroids and the influence of rootstocks on field performance of Navel orange. *Proc.11th Conf. IOCV* (eds. RH Brlansky, RF Lee, and LW Timmer). Univ. Calif., Riverside, CA, pp. 234–239.

Roistacher CN, Bash JA, and Semancik JS (1993). Distinct disease symptoms in *Poncirus trifoliata* induced by three viroids from three specific groups. *Proc.12th Conf. IOCV* (eds. P Moreno, JV daGraca, and LW Timmer). Univ. Calif., Riverside, CA, pp.173–179.

Roistacher CN, Calavan EC, Nauer EM, and Bitters WP (1979). Spread of seedling yellows tristeza at Research Center. *Citrograph* 64: 167–169.

Roistacher CN, Nauer EM, Kishaba A and Calavan EC (1980). Transmission of *Citrus* tristeza virus by *Aphis gossypii* reflecting changes in virus transmissibility in California. *Proc. 8th Conf. IOCV* (eds. EC Calavan, SM Garsney, and LW Timmer). Univ. Calif., Riverside, CA, pp. 76–82.

Roistacher CN, Calavan EC, Blue RL, Navarro L and Gonzales R (1977). A new more sensitive citron indicator for detection of mild isolates of citrus exocortis viroid (CEV). *Plant Disease Rept.* 61: 135–139.

Rustici G, Accotto GP, Norris E, Masenga V, Luisoni E, and Milne RG (2001). Indian *Citrus* ringspot virus: a proposed new species with some affinities to Potex, Carla, Fovea and Allexi viruses. *Archives of Virology* 145 (9): 1895–1908.

Salibe AA (1980). Further studies on exocortis disease of citrus. *Proc. 8th Conf. IOCV* (eds. EC Calavan, SM Garnsey, and LW Timmer). Univ. Calif., Riverside, CA, pp. 215–219.

Sano T, Hataya T, and Shikata E (1988). Complete nucleotide sequence of a viroid isolated from Etrog citron, a new member of hop stunt group. *Nucleic Acid Res.* 16: 347.

Satyanarayana T, Bar-Joseph M, Mawassi M, Albiach-Marti MR, Ayllon MA, Gowda S, and Hilf ME (2001). Amplification of *Citrus* tristeza virus from a cRNA clone and infection of citrus trees. *Virology* 280 (1): 87–96.

Semancik JS and Weathers LG (1968). Exocortis virus of *Citrus* association of infectivity with nucleic acid preparations. *Virology* 36: 326–328.

Semancik JS and Weathers LG (1971). Exocortis virus—an infectious free nucleic acid plant virus with unusual properties. *Virology* 47: 456–466.

Semancik JS and Weathers LG (1972). Exocortis disease: evidence for a new species of "infectious" low molecular weight RNA in plants. *Nature New Biol.* 27: 242–244.

Semancik JS and Duran-Vila N (1991). The grouping of *Citrus* viroids: additional physical and biological determinants and relationship with diseases of citrus. *Proc.11th Conf.*

IOCV (eds. RH Brlansky, RF Lee, and LW Timmer). Univ. Calif., Riverside, CA, pp. 178–188.

Semorile LC, Dewey RA, Gracia ML, DalBo E, Ghiringhelli PF, Romanowski V, and Grau O (1993). CDNA clones of CTV that discriminate severe and mild strains. *Proc.12th Conf. IOCV* (eds. P Moreno, JV daGraca, and LW Timmer). Univ. Calif., Riverside, CA, pp. 28–32

Shikata E (1990). New viroids from Japan. *Seminars in Virology*, vol. 1, pp. 107–115

Scoric D, Conerly M, Szychowski JA, and Semancik JS (2001). CEVd-induced symptom modification as respose to a host-specific temperature-sensitive reaction. *Virology* 280 (1): 115–123.

Tanaka H and Imada J (1974). Mechanical transmission of viruses of satsuma dwarf, *Citrus* mosaic, Navel infectious mottling, and Natsudaidai dwarf to herbaceous plants. *Proc. 6th Conf.* (eds. LG Weathers and M Cohen) Univ. Calif., Div. Agric. Sciences, pp. 141–146.

Terrada E, Kerschbaumer EJ, Guinta G, Galeffi P, Rimmler G, and Cambra M (2000). Fully "Recombinant enzyme-linked immunosorbent assays" using genetically engineered single chain antibody fusion proteins for detection of *Citrus* tristeza virus. *Phytopathology* 90 (12): 1337–1344.

Thornton IR, Emmett RW, and Stubbs LL (1980). A further report on the grapefruit tristeza pre-immunization trial at Mildura. *Proc. 8th Conf. IOCV* (eds. EC Calavan, SM Gransey, LW Timmer). Univ. Calif., Riverside, CA, pp. 51–53.

Tsai Mei-Chen, Su Hong-Ji, and Garnsey SM (1993). Comparative study on stempiting strains of CTV in the Asian countries. *Proc. 12th Conf. IOCV* (eds. P Moreno, JV DaGraca, and LW Timmer). Univ. Calif., Riverside, CA, pp. 16–19.

Tsuchizaki T, Sasaki A, Saito Y (1978). Purification of *Citrus* tristeza virus from diseased *Citrus* fruits and the detection of the virus in *Citrus* tissues by fluoresent antibody techniques. *Phytopathology* 68: 139–142.

Vela C, Cambra M, Sanz A, and Moreno P (1988). Use of specific monoclonal antibodies for diagnosis of *Citrus* tristeza virus. *Proc.10th Conf. IOCV* (eds. LW Timmer, SM Garnsey, and L Navarro). Univ. Calif., Riverside, CA, pp. 55–61.

Vives MC, Galipienso L, Navarro L, Moreno P, and Guerri J (2001).The nucleotide sequence and genomic organization of *Citrus* leaf blotch virus: Candidate type species for a new virus genus. *Virology* 287: 225–233

Vogel R and Bove JM (1968). Cristacortis, a virus disease inducing stempitting on sour orange and other *Citrus* species. *Proc 4th Conf. IOCV* (ed. JFL Childs). Univ. Florida Press, Gainesville, Fla, pp. 221–228.

Vogel R and Bove JM (1972). Relationship of cristacortis virus to other *Citrus* viruses. *Proc. 5th Conf. IOCV* (ed. WC Price). Univ. Florida Press, Gainesville, Fla, pp.178–184.

Vogel R and Bove JM (1976). Evidence for the existence of strains of cristacortis pathogen. *Proc. 7th Conf. IOCV* (ed. EC Calavan). Univ. Calif., Riverside, CA, pp. 123–136.

Vogel R and Bove JM (1980). Pollen tansmission to citrus of the agent inducing cristacortis and psorosis young leaf symptoms. *Proc. 8th Conf. IOCV* (ed. EC Calavan). Univ. Calif., Riverside, CA, pp. 188–190.

Wallace JM (1957).Tristeza and seedling yellows of *Citrus* plants. *Plant Disease Reptr.* 41: 394–397.

Wallace JM and Drake RJ (1962). Tatter leaf, a previously undescribed virus effect on citrus. *Plant Dis. Reptr.* 46: 211–212.

Wallace JM, Martinez AL and Drake RJ (1965). Further studies on *Citrus* seedling yellows. *Proc. 3rd Conf. IOCV* (ed. WC Price). Univ. Florida Press, Gainesville, Fla, pp. 36–39.

Yamada S and Sawamura K (1952). Studies on the dwarf disease of Satsuma orange, *Citrus unshiu* Marcovitch. (Preliminary report) *Bull. Hort. Div. Tokai-Kinki Agr. Exp. Stn.* 1: 61–71.

Yang ZhongNan, Ye XingRong, Choi Sandong, Molina J, Moreno F, Wing RA, et al. (2001). Construction of 1.2 Mb contig including the *Citrus* virus resistance gene locus using a bacterial artificial chromosome library of *Poncirus trifoliata* (L.). *Genome* 44 (3): 382–393.

Zhou Changyong, Xueyuan Zhao, and Yuanhui Jiang (1993). Goutoucheng—a new indicator plant for citrus mosaic virus. *Proc. 12th Conf. IOCV* (eds. P Moreno, JV da Graca, and LW Timmer). Univ. Calif., Riverside, CA, pp. 368–370.

Zhou Changyong, Xueyuan Zhao, Yuanhui Jiang, and Xinhua He (1993). The occurrence of Satsuma dwarf virus in China. *Proc.12th Conf. IOCV* (eds. P. Moreno, JV daGraca, and LW Timmer). Univ. Calif., Riverside, CA, pp. 349–351.

Nematode Diseases

Ashkenazi S and Oren Y (1977). Rootstocks, scions and soil fumigation in replanting 'Shamouti' orange. *Proc. Int. Soc. Citriculture,* vol. 2, pp. 638–639.

Badra T and Elgindi DA (1979). The relationship between phenolic content and *Tylenchulus semipenetrans* populations in nitrogen amended *Citrus* plants. *Revue de Nematologie* 2: 161–164.

Baines RC, Clarke OF, and Bitters WP (1948). Susceptibility of some *Citrus* species and other plants to the *Citrus* root nematode *Tylenchulus semipenetrans. Phytopath.* 38: 912 (Abstr).

Basile M, Lo Giucide, and Inserra RN (1977). Problemi connessi al reimpianto degli agrumenti in Sicilica. 1) Trattamenti nematocidi a base di bromuro di metile e residui di bromo nel terreno. *Nematol. Medit.* 5: 57–64.

Byars LP (1921). Notes on the *Citrus* root nematode *Tylenchulus semipenetrans* Cobb. *Phytopath.* 2: 90–94.

Cobb NA (1913). Notes on *Monochus* and *Tylenchulus. J. Wash. Acad. Sci* 3: 287–288.

Cohen E (1966). The development of *Citrus* nematode on some of its hosts. *Nematologica* 11: 593–600.

Elgindi AY, Ahmed SS, and Oteifa BA (1976). Effects of nonfumigant nematicides on root populations and manganese and zinc levels in rough lemon seedlings infected with the *Citrus* nematode *Tylenchulus semipenetrans. Plant Dis. Reptr.* 60 (8): 682–683.

Fattah FA, Saleh HM, and Aboud HM (1989). Parasitism of *Citrus* nematode *Tylenchulus semipenetrans* by *Pasteuria penetrans* in Iraq. *J. Nematology* 21: 431–433.

Kalyviotis-Gazelas C and Koliopanos CN (1972). Control of the *Citrus* nematode (*Tylenchulus semipenetrans* Cobb 1913) with DPCP and Thionazin. *Ann. Inst. Phytopathol. Benaki.* 10: 229–235.

Gaspard JT and Mankau R (1986). Nematophagous fungi associated with *Tylenchulus semipenetrans* in the *Citrus* rhizosphere. *Nematologica* 32: 359–363.

Koliopanos CN et al. (1976). Results of dibromo chloropropane treatment on the *Citrus* nematode *Tylenchulus semipenetrans* Cobb 1913) on orange trees. *Ann. Inst. Phytopathol. Benaki.* 11: 176–186.

Koliopanos CN et al. (1979). Control of *Tylenchulus semipenetrans* Cobb 1913 with various nematicides. *Ann. Inst. Phytopathol. Benaki* 12: 72–80.

Lo Giudice V (1981). Present status of *Citrus* nematode control in Mediterranean area. *Proc. Int. Soc. Citriculture* vol. 1, pp. 384–386.

Mankau R (1980). Biological control of nematode pests by natural enemies. *Ann. Rev. Phytopathol.* 18: 415–440.

Poucher C, Ford HW, Suit RF, and Du Charme EP (1967). Burrowing nematode of *Citrus.* Fla. Dept. Agric. Divn. Plant Ind. Bull. 7, 63 pp.

Roccuzzo G, Ciancio A, and Lo Giudice V (1992). Some observations on the Ecology of *Citrus* Nematode *Tylenchulus semipenetrans* Cobb in Southern Italy. *Proc. Int. Citriculture,* vol. 3, pp. 950–952.

Salem AAM (1982). Observation on the population dynamics of the *Citrus T. semipenetrans* in Shark ea Govemote. *Egyptian J. Phytopath.* 17 (1/2): 31–34.

Tarjan AC (1964). Plant parasitic nematodes in the Unted Arab Republic. *FAO Pl. Prot. Bull* 12: 49–56.

Thapa Anjana (1991). Study of nematodes problem associated with *Citrus* dieback in Darjeeling district of West Bengal. Ph.D. thesis, Bidhan Chandra Krishi Vishwavidyalaya, West Bengal.

Thapa Anjana and Mukhopadhyay S (1994). Studies on the problems of *Citrus* nematodes in orchards and nurseries of mandarin orange in Darjeeling hills and their management. *Indian J. Nematol.* 24 (2): 214–220.

Van Gundy SD and Garabedian S (1981). Alternatives to DBCP for *Citrus* nematode control. *Proc. Int. Soc. Citriculture,* vol. 1, pp. 387–390.

Vilardebo A, and Luc M (1961). Le "slow decline" des *Citrus* du nematode *Tylenchulus semipenetrans* Cobb. *Fruits* 16: 445–454.

Vilardebo A, Squalli, and Devaux R (1975). Utilisation possible du DBCD, du Phenamiphos et du Prophos contre *Tylenchulus semipenetrans* dans les vergers du Marco. *Fruits* 30: 317–327.

Vovlas N, Lamberti F, and Inserra RN (1973). Results of glasshouse experiments with new nematicides against the *Citrus* nematode *Tylenchulus semipenetrans* Cobb. *Int. Citrus Congress Murcia-Valencia* 2: 687–691.

Walter DE and Kaplan DT (1990). Antagonists of plant parasitic nematodes in Florida citrus. *J. Nematology* 22: 567–573.

8

Pests and Their Management

More than 875 species of insects and mites may occur on different species of *Citrus* throughout the world (McLaren, 1978). Insects, like pathogens, were inherent components of the citrus ecosystem in the primary forests of the ancient world. With commencement of domestication of different citrus species, their introduction into different countries, varietal improvement and large-scale cultivation through application of chemical technologies, caused some of these insects to become severe pests, destroying their mutual existence with predators and parasites and disrupting ecological harmony. Insects with high powers of dispersal and rapid reproductive rates have the greater potentiality for becoming severe pests.

DISTRIBUTION OF CITRUS PESTS IN VARIOUS COUNTRIES

The distribution of *Citrus* pests in various countries is not uniform. It varies according to the ecological characters of the concerned region. In the same region again their distribution varies according to the phenology of the plants and the season of observation. The *Citrus* pests that occur worldwide irrespective of the ecological conditions and the season are given in Appendix XII.

Major pests of citrus are (1) Thrips, (2) Aphids, (3) Psyllids, (4) Whiteflies, (5) Scales, (7) Mealybugs, (8) Leaf miners, (9) Butterfly, (10) Fruit fly, and (11) Mites. Occasional incidences of borers, weevils, snails, ants, leafhoppers, and midges have allso been reported from some countries.

THRIPS

Thrips are common pests of citrus in various countries. Three species of thrips generally infests *Citrus, Heliothrips haemorroides* Bouche, *Scirtothrips citri* (Moulton) and *Scirtothrips aurantii* Faure—depending on the geographic and ecological conditions of the region. *S. citri* is a major pest in California whereas *S. aurantii* is a serious pest of South

Africa. *H. haemorroides* is common in Mediterranean climate. Benfatto et al. (2000) reported a new species, *Pezothrips kellyanus* (Bagnall), from Sicily.

Symptoms

Appearance of symptoms depends upon the degree of attack and the growth stage of the plant. Infestation normally occurs on the leaves and fruits. Severe infestation may cause stunted growth and excessive drop of injured fruitlets. There may be scribbling on foliage. Young leaves and stems become thickened, distorted; shoot apex turns black, dies and falls off. Injury to flush foliage is characterized by two thickened whitish-gray streaks on either side of the midrib on the upper leaf surface, and misshapen and curled leaf margins. Infestation of fruitlets that are at the stemend causes stemend blemish. Feeding punctures in the rind epidermis causes stemend ring scarring. Second generation infestation causes culled fruits showing extensive scabby, scarified, grayish tissue around the shoulder and stylar end of the fruit. *P. kellyanus* causes ringlike symptoms of damage to lemons and oranges.

Thrips also infest mango trees and may cause lesions on mango fruit, deformation of leaves, and stunted plant.

Biology

With favorable food and weather, a female may lay 200–250 eggs. These are inserted by a serrated ovipositor beneath the cuticle of new foliage and, to a lesser extent, in green stems, twigs, buds, and small fruits. There are two immature feeding stages, the first and second instars, followed by two resting (nonfeeding) stages, the prepupa and pupa, which search for cracks or crevices within the tree to hide in or drop to the ground during pupation. Adults are winged, active fliers and may disperse to nearby hosts (Morse, 1986).

Citrus thrips overwinter in the egg stage. The eggs remain dormant until an adequate number of thermal heat units is accumulated, then eclosion takes place. The three feeding stages (first and second instar nymphs and the adult) prefer to feed on tender and young leaves. They may complete 8–12 generations in a year (Tanigoshi, 1981).

The thermal threshold for egg development is probably below 18.3° C. The upper temperature limit is approximately 37.8° C.

Monitoring and Forecasting

Monitoring and surveillance of thrips is generally done by trapping the adults at the stage of their emergence from the pupae. Second stage larvae normally drop to the ground cover debris to pupate. Adults emerging from the pupae migrate back to the tree's canopy or other hosts. Tanigoshi and Moreno (1981) fabricated a trap from PVC pipe 10 cm high and inside diameter of 20.3 cm. A 439 cm^2 clear acetate plate evenly coated on both sides with Tangelfoot and weighing 18 g was placed over the top opening of the trap. The trap was set under the tree canopy to catch the insects. Each plate was then covered in clean venyl folders (645 cm^2) in the field to minimize contact with the Tangelfoot. The plates were then stacked and stored in the refrigerator for subsequent detailed examination under a stereomicroscope. Samways et al. (1987) introduced a fluorescent yellow, one-sided trap as a means for monitoring citrus thrips. Grout and Richards (1992) introduced a smaller double-sided trap consisting of nonfluorescent yellow polyvinyl chloride. The trap was 140 mm × 76 mm on both sides. Reverfly was used as an adhesive coat and applied to the traps either with a paintbrush or by dipping the cards in 50% water solution of Reverfly. The traps were exposed for a week, then removed and covered with a thin layer of plastic (polyethylene) before removal to the laboratory for counting.

Grove et al. (2000) tested the efficacy of colored sticky traps for *S. aurantii* in South Africa. According to them, yellow traps can be used effectively for assessing activity level of *S. auarantii* in mango orchards. Conti et al. (2001), on the contrary, found white traps more successful in monitoring the population in citrus fields.

To forecast the incidence of larvae/adults, Tanigoshi (1981) developed a multivariate model driven by temperature, whose major state variables were day-degrees, accumulated citrus-thrips days, citrus tree phenology (budding, bloom, petal-fall, flush), incremental fruit growth rates, and natural enemy populations. This model was useful in determining the time of application of chemical treatment if required.

Natural Enemies

Records on natural enemies of thrips are very limited. Tanigoshi et al. (1984, 1985) indicated that application of predaceous phytoseiid mite *Euseius addoensis* (McMurty) (Aceri: Phytoseiidae) is effective in biological control of thrips. *Amblyseius degenerans* (*Iphiseius degenerans*) mite also feed on thrips. Inamullah-Khan and Morse (1999) observed the predatory effect of two species of *Chrysoperla* mite species, *C. carnea* and *C. rufilabris*. According to them augmentative release of these species has potential for suppressing citrus thrips levels in California.

Integrated Management

Before emergence of the concept of the integrated management, chemicals used to be recommended to control the infestation of thrips. Some of the recommended sprays were 0.04% a.i. of temephos at 100% petal fall or double spray at 0.01% a.i. at 100% petal fall, with a second spraying 5–6 weeks later. But this chemical adversely affects the natural enemies of thrips, the scale insects. Single application of dimethoate 5G or 40% EC in basin irrigation water at 20% petalfall was also in use. Subsequent recommendations included the use of formetanate (carzol), dimethoate (cygon), rynia + sugar (Ryan 50) bait, acephate (Orthene), and several pyrethroids. But these chemicals may cause rapid reinfestation after the treatment, given and the propensity of thrips to rapidly develop resistance to them (Morse and Brawner, 1986). Conti et al. (2001) evaluated the efficacy of different chemicals and biopesticides and observed the IPM compatibility of abamectin and lufenuron.

Grout and Richards (1992) recommended integrated management introducing physical barriers to the flight of thrips to the field, monitoring the thrip population during the vulnerable period, and application of safe chemical treatments as and when necessary. They observed *Grevillea robusta* A. Cunn to be a host of *Citrus* thrips. When this plant was used as windbreak, pest damage increased. *Casurina cunninghaminiana* Miq. on the other hand, planted as windbreak, produced a food source for predaceous mite populations for natural biological control of incoming thrips. Thus by planting proper windbreaks and monitoring of the threshold population of thrips in the citrus canopy, application of disruptive chemical measures can be reduced.

APHIDS

Five species of aphids are generally considered of quarantine significance: *Aphis citricola* (Green citrus aphid), *Aphis gossypii* (Cotton aphid), *Myzus persicae* (Green peach aphid), *Toxoptera aurantii* (Black citrus aphid) and *Toxoptera citricidus* (Brown citrus aphid). Another aphid, called *Sinomeougoura citricola* Vander Gat. feeds on *Citrus* and often becomes a serious pests. Aphids are more important as vectors of viruses.

Symptoms

Aphids colonize on the fully expanded and hardened leaf flush covering veins and petioles. In severe infestation, leaves become deformed, curled and pale yellow; young shoots may become twisted and the growing apex may wither.

Biology and Ecology

Aphids are piercing insects with sucking mouthparts. They mostly feed on phloem tissue. The characteristic features of an aphid are its pear shape and propagation of two forms—winged and wingless. The latter form predominates in a population. Eggs laid in autumn overwinter and give rise to apterous viviparous parthenogenetic females in spring. During summer several generations are produced. Only apterous females appear in the first or second generations that are parthenogenetic, viviparous and live on primary hosts. Winged parthenogenetic, viviparous females or migrants arise in the second, third and later generations. They may live on primary hosts for some time, then fly to secondary or alternative hosts for overwintering or even invade fresh crops. On alternative hosts they may produce males and females and undergo sexual reproduction. Black citrus aphids have a wide host range but brown citrus aphids feed only on *Citrus* and the relatives. In brown citrus aphids the adult has a characteristic brown spot on the dorsal margin of its body. When adults of the two species are squeezed, the blood of the black species appears red while that of the brown species appears yellow.

Aphid incidence is mostly seasonal. Gurung et al. (1993) surveyed the incidence of citrus aphids in Darjeeling hills (26° 25′ and 27° 10′ N; 88° 10′ and 88° 50′ E). Distribution was found to be different both with respect to species and season. The prevalent species are *T. citricidus, T. aurantii* and *S. citricola. A. citricola* and *A. spiricola* are more or less location specific whereas *S. citricola, T. aurantii* and *T. citricidus* are found in all locations. The population of these insects becomes high during August-September. *T. aurantii* and *T. citricidus* are also quite dominant during March–April but scarce in November–December when the average maximum and minimum temperatures decrease from 24.4 to19–9° C and 15.4 to 8.8° C respectively. Higher trapping of alates during this period indicates their migration to a suitable environment or overwintering.

Natural Enemies and Biological and Integrated Management

Natural enemies of aphids are generally aphid specific. Among citrus aphids, extensive work has been done on the vector species *T. citricidus* that occurs almost throughout the world, notably Asia, Australia, New Zealand, Pacific islands, Africa and South America. It has also spread to Central America and the Caribbean islands, however, the current trend shows its gradual spread to Middle America and main land, USA.

There are several predators, parasites, and entomopathogens that prey on *T. citricidus*. Predators generally arrive when aphids are rapidly increasing but their number lags behind that of aphids. The typical pattern is that predators move in when aphids are abundant and arrive too late or too early to be true regulators of the aphid population. Yokomi (1992) has listed the potential predators, parasites, and entopathogens (Appendix XIII). If proper ones are identified and delivered in proper time, they may effectively control the aphid population.

Essential aspects of biological and integrated control of vector aphids have already been presented in Chapter 7 dealing with "Diseases".

PSYLLIDS

Two species of psyllids are of quarantine importance: *Diaphorina citri* (Asian citrus psyllid) and *Trioza erytreae* (South African citrus psyllid). Usually they do not per se cause much damage to *Citrus* as pests. They are important mostly as vectors of the "Greening disease". *D. citri* though common in Asian countries, has also been reported from Brazil (Aubert and Xia, 1990) and Venezuela (Cerameli et al., 2000). Similarly, *T. erytreae,* though common in South Africa has been recorded across the African continent and in the Arabian peninsula, as well as in various neighboring islands (Samways, 1990). Major distributional differences between the two species are the colonization of the entire forested area south of the Sahara by *T. erytreae* beyond a certain altitude threshold, while the natural habitat of *D. citri* does not usually include Indo-Malayan forests.

Symptoms

In acute infestation, *D. citri* may cause twisting of the leaves and *T. erytreae* formation of characteristic bumps on the undersurface of leaves.

Biology and Ecology

An adult psyllid is brown and may be 2.4 mm long. It mostly remains on the undersurface of leaves or on young shoots and feeds in an angular position with the head touching the surface and the rest of the body upraised. Eggs remain in the folds of half-opened leaves and axils of leaves or pushed in between buds and stems or petioles of leaves or axillary buds. One female may lay 800 eggs. Oviposition period continues about two months. Incubation is 4–6 days in summer and up to 22 days in winter. There are five nymphal instars. Nymphal period about 11–25 days. Total life cycle 15–47 days depending on season. Adult longevity 90–189 days depending on the climate, season, and species. There may be 9 generations in a year.

Trioza erytreae does not establish freely in arid or semiarid climate with low rainfall and high temperature but occurs in subtropical climates with an altitude of 1,000–2,000 m (Aubert, 1987). *D. citri,* on the other hand, is more tolerant to climatic extremes. It is found from sea level up to an elevation of 1,500 m (Lama et al., 1988). The host range of both species is restricted to a few related host plants. Egg-laying and nymphal development of *T. erytreae* is restricted to 15 and 13 species respectively whereas breeding hosts of *D. citri* are restricted to 21 species of the family Rutaceae.

Natural Enemies, Biocontrol and Integrated Management

In nature, several predators and parasites pray on psyllids in the Asian and Pacific region. Among the predators, the common ones are *Chilocoris nigritis* Fabr., *Coccinella septempunctata* L., *C. rependa* Thump, *Menochilus sexmaculatus* Fabr., and *Brumus suturalis* Febr. Parasites include 15 species of Hymenopteran insects known to be associated with *D. citri,* belonging to five families of Chalcidoidae: Eulophidae (2 species), Encyrtdae (> 6 species), Signiphoridae (1 species), Pteromalidae (1 species), and Aphelinidae (5 species) (Appendix XIV, Qing, 1990). Among these, only two species, an ectoparasite, *Tamarixia radiata* (Waterson) and an endoparasite *Diaphorecyrtus aligarhensis* (Shafee, Alam, and Agarwal), have been recorded as primary parasites; remaining 13 species acting as secondary or tertiary hyperparasites. *Tamarixia radiata* has been successfully applied to control *D. citri* on Reunion Island (Aubert, 1987).

Biocontrol and integrated management of psyllids have been properly dealt with in Chapter 7, "Diseases".

WHITE FLIES

Four whiteflies are of quarantine significance: *Aleurocanthus woglumi* (Citrus Black fly), *Aleurothrixus floccosus* (Woolly Whitefly), *Dialeurodes citri* (Citrus Whitefly), and *Dialeurodes citrifolii* (Cloudy-wing Whitefly). Among these, Black fly and Citrus whitefly often assume the status of pests in some countries. Recently Chen-BingXu et al. (1999) reported a new species *Aleurocanthus spiniferus* (Citrus spiny whitefly) feeding on citrus in China. Luo Zhi Yi et al. (2001) from the same country described 11 species of citrus whitefly, namely, *Aleurolobus subrotundus, Aleurotuberculatus aucubae, A. jasmini, Bemisia afer, B. giffardi, B. tabaci, Dialeurodes citri, D. citricola, D. citrifolii, D. kirkaldyi,* and *Parabemisia myriae.* Considering the international significance of the black fly and citrus whitefly only these two have been described here.

Citrus black fly (*Aleurocanthus woglumi* Ashmed: Aleurodoida: Homoptera)

Citrus black fly normally feeds on leaves. Severe infestation causes damage to leaves and often the shoot apex. But the most important feature of the fly is the production of honeydew that contributes to the severe accumulation of sooty mold fungus. The fly is small and inconspicuous. Its distribution was earlier primarily restricted to the tropics and subtropics but it has now spread to citrus orchards of many temperate countries.

Biology and Ecology

The fly lays eggs in spirals with three whorls. One spiral contains 15–22 eggs. One female normally lays three spirals during her life. Incubation period varies from 7–14 days. Nymphal period varies from 38–60 days depending on the season. Pupation varies from 100–131 days in summer and 147–161 days in winter. Emergence is a day-degree dependent process.

Adult fly dark orange with smoky wings and forewings have four whitish areas of irregular shape. There are two distinct broods of this insect in a year.

Natural Enemies and Biological Control

Knapp and Browning (1989) illustrated black fly management in citrus orchards. Use of natural enemies forms an important component of this management. Several predators and parasites feed on this insect. Common predators are *Malada boninensis* (Chrysopidae: Neuroptera) and *Brumus suturalis* (Coccinellidae: Cleopetra). But the natural fecundity of predators is not high enough to compensate the natural fecundity of the flies.

Parasites are mostly hymenopteran insects: *Prosopaltella divergens, Encarsia merceti, Eretmocerus serius, Amitus hesperidium* Silv., *Encasia opulenta* Silv., *E. clypeali, and Prosopaltella icornensis* Howard. Among these, the last four parasitoids are very effective in controlling black flies.

Several fungi are also effective against black fly: *Aschersonia aleyrodis, A. papillata, A. goldiana, Aegirita webberi, Verticillium lecanii,* and *Beauveria basiana.* The last fungus is a canddidate for commercial application.

Citrus Whitefly (*Dialeurodes citri* Ashmed)

Biology

This fly lays eggs singly, irregularly scattered on the undersurface of leaves. One female may lay 200 eggs. Oviposition continues for about 4 days. Incubation varies from 10–12 days in the first brood and 7–13 days in the second brood. There are 3 nymphal and 1 pupal instars. Duration of the nymphal period is seasonal, normally 25–51 days in summer and 114–159 days in winter. The preimaginal period is 177–190 days for the summer and 181–205 days for the winter brood. Ecological features and management practices are more or less similar to those of black fly. Argov et al. (1999) monitored the citrus whitefly in *Citrus* orchards in Israel to undertake control decisions in the IPM before establishment of natural enemies in view of the fact that the main damage is caused by sooty mold colonizing the honeydew of the pest. Using Taylor's Power Law to describe the mean variance relationship, they found the dispersion of citrus whitefly to be aggregated at all developmental stages. They estimated the action threshold using regression of sooty mold level on egg or larval counts. Two-stage sample plans were constructed in different pest generations for the egg and larval stages, on the basis of variance between leaves of a tree and variance between trees in a plot. They prepared fixed sample size plans in order to attain a given precision level. Sequential sample plans were also drawn. Simulation of the sequential sampling plans showed them to be highly reliable.

SCALE INSECTS

Scale insects very widely infest different species of *Citrus*. Numerically they are the largest group of insects affecting this crop. The members mostly belong to families Diaspididae, Coccidae and Margarodidae. They are normally known by different common names (Table 8.1).

Watanabe et al. (2000) surveyed the *Citrus* orchards in Brazil and found the following armored scales occurring almost throughout the year: *Selenaspidus articulatus* (dominant species), *Parlatoria ziziphi*, *Mycetapsis personata*, *Cornuapis beckii*, and *Chrysomphalus ficus* (*Chrysomphalus aonidum*). 10 scales per leaf was the damage threshold level.

General Features and Biology

Scale insects are normally sessile, small and often inconspicuous. Some scale insects may be mobile during all stages of their development. Several insects may become immobile after the crawler stage and remain soft (non-armored scales, Coccidae) or bear armor (armored scales, Diaspididae). Adult females may be flat, circular or elongate-ovate, with integument smooth or covered with wax. They may or may not be legged. Antennae may be present or reduced.

A gravid female lays a few to several thousand eggs, which are protected under cover of its body. Most males are winged and short-lived. But reproduction is parthenogenetic in many insects. Depending on the species and prevailing temperature, there may be 1–10 generations per year. The number of generations is normally high in tropical climates.

Wakgari and Giliomee (2000) studied the fecundity, body size, and phenology of *Ceroplastes destructor* Newstead. They observed significant variation between female size classes. Female body size is significantly positively correlated with fecundity. Larger individuals have a longer oviposition period. In South Africa, decline of the population density of the second instar takes place in February but the population of the third instar nymphs steadily increases and extends to the end of July. The peak population of the adults is found in August.

Table 8.1: Common and scientific names of scale insects belonging to different families

Serial No.	Common name	Scientific name	Family
1.	California red scale	*Aonidiella aurantii* (Maskell)	Diaspididae
2.	Yellow scale	*Aonidiella citrina* (Coquillat)	Diaspididae
3.	Purple scale	*Chrysomphalus ficus* Ashmead	Diaspididae
		Lepidosaphes beckii (Newman)	Diaspididae
4.	Glover scale	*Lepidosaphes (= Insulapes) gloverii* (Packard.)	Diaspididae
5.	Black Parletoria scale	*Parletoia zizyphus* (Lucas)	Diaspididae
6.	Chaff scale	*Parletoria pergandii* Comstock	Diaspididae
7.	Dictyospermum (fern) scale/ (Spanish red scale)	*Chrysamphalus dictyospermi* (Morg.)	Diaspididae
8.	Aspidistra scale	*Pinnaspis aspidisrae* (Sign.)	Diaspididae
9.	Mussel scale	*Mytilococcus beckii* (Newman)	Diaspididae
10.	Citrus long scale	*Mytilococcus gloverii* (Pachard)	Diaspididae
11.	Citrus snow scale	*Unaspis citri* (Comstock)	Diaspididae
12.	Arrowhead scale	*Unaspis yanonensis* (Kuwana)	Diaspididae
13.	Rutous scale	*Selenaspis articulatus* Morgan	Diaspididae
14.	Cottony cushion scale	*Icerya purchasi* Maskell.	Margarodidae
15.	Mediterranean black scale	*Saissetia oleae* (Oliver)	Coccidae
16.	Hard brown scale/ Coffee helmet scale	*Saissetia hemisphaerica* Targ	Coccidae
17.	Brown soft scale	*Coccus hesperidum* L.	Coccidae
18.	Citricola scale	*Coccus pseudomagnoliarum* (Kuwan)	Coccidae
19.	Coffee green scale	*Coccus viridis* Green	Coccidae
20.	Florida white scale	*Ceroplastes floridensis* Comstock	Coccidae
21.	White wax scale	*Ceroplastes destructor* Newstead	Coccidae
		Pulvinari floccifera Westwood	Coccidae
22.	Pink wax scale	*Ceroplastes rubens* Maskell	Coccidae
23.	Chinese wax scale	*Ceroplastes sinensis* Del Guer.	Coccidae
24.	Fig wax scale	*Ceroplastes rusci* L.	Coccidae
25.	Black scale	*Chrysomphalus aonidum*	Diaspididae

There may be three stages in a female from crawler to adult. The long stylet of the mouth-parts of all stages can be found in plant tissues below the resting adults. Efficient scales can penetrate to phloem cells and feed for a long time. Copious waxy deposits usually cover the body of the non-armored scales and dense colonies may develop on branches and trunks of the host plants.

Predators and Parasites

Occurrence of predators and parasites together with the scale insects has been a natural phenomenon on citrus plants from very ancient times. It became a matter of concern when cottony cushion scale-infested citrus germplasms were introduced from Australia into California in 1868 and predators and parasites of scales native to Californian citrus were found nonfeeders of those scales. When the predatory (vedalia beetle, *Rodolia cordinalis* (Mulsant)), naturally occurring on Australian citrus, was introduced to Californian citrus, it fed voraciously on the imported scales. Subsequently intensive and extensive search for predators and parasites of various scale insects was indertaken in most of the citrus-growing countries of the world and a large number recorded (Argyrou and Mourikis, 1981) and given in Table 8.2.

Wakgari (2001) identified seven primary and three secondary parasitoids of *Ceroplastes destructor.* Of the total primary parasitoids, *Aprostocetus* (= *Tetrastichus*) *ceroplastae* constituted 78%. Generally the appearance of peak numbers of parasitoid and predator synchronizes with the peak emergence of scale adults. The parasitoid and predator association with the scale contain a density dependent regulatory mechanism. Parasitoids demonstrate density-dependence at the third instar and preovipositing female stage depending on the host species, i.e. the third instar of *Citrus reticulata* and preovipositing stage of *S. malaccansis.* Predators, on the other hand, show a density-dependent mortality factor only during the preovipositing female stage of the scale *C. reticulata.* The density-dependent process of parasitism and predation fluctuates between generations. Wakgari and Giliomee (2000) conducted key factor analysis to determine and quantify the contribution of individual mortality factor vis-à-vis total generation mortality. Key stage mortality determined from a cohort life table is third instar and preovipositing female. Parasitoids, predators, and miscellaneous factors primarily cause the mortality of *C. destructor.* Parasitoids and miscellaneous factors act as density dependent regulatory agents during preovipositional and first instar stages respectively. Some of the mortality factors act randomly with no reference to population densities or in an inverse density-dependent manner, during the egg crawler second or third instar stages.

Management

Chemical treatment remained the reliable means of controlling scales until the 1970s or even 1980s. Chemicals in use, includes low range petroleum oil, malathion and oil, carbaryl and/or oil, azimphos methyl and/or oil, methidathion and or oil, parathion and/or malathion, chloropyrifos, etc. (Morse, 1986). But application of these chemicals adversely affects the plants and fruits in various ways. Oils affect photosynthesis, transpiration, and respiration. Parathion tends to accelerate yellowing, drop of older leaves, and applications preceding flush of growth can delay the appearance and development of new leaf tissues. In general, parathion and ethion cause marginal visible injury to *Citrus* trees. Carbaryl, malathion, methidathion and azimphos methyl, buprofezin, pirimiphos methyl, etc. apparently cause no observable phytotoxic response. But

Table 8.2: Predators and parasites of scale insects

Species	Predators	Parasites
I. Diaspididae		
1. *Aonidiella aurantii* (California red scale)	*Aphytis melinus** DeBach *A. chrysomphali* Mercet	*Chilocorus bipustulatus* L. *Exochomus quadripustulatus* L. *Lindoruslophanthae* Blaist
2. *Lepidosaphes beckii* (purple scale)	*Aphytis lepidosaphes** Compere *Aspideotiphagus citrinus* Craw	*C. bipustulatus* *E. quadripustulatus*, L. *lopanthae*
3. *Aspidiotus nerii*	*Aphytis chilensis** Howard, *A. chrysomphali* Mercet, *A. melinus* DeBach, *Aspidiotiphagus citrinus* Craw	*C. bipustulatus, E. quadripustulatus, L. lophanthae*
4. *Perlatoria zizyphus* (black perlatoria scale)	*Aspidiotiphagus lounsburyi* Berlesi et Paoli	
5. *Chrysomphalus dictyospermi* Dictyospermum (fern) scale	*Aphytis chrysomphali, A. melinus**, *Aspidiotiphagus citrinus*	*C. bipustulatus, E. quadripustulatus, L. lophanthae*
6. *Chrysomphalus aonidum* (black scale)	*Aphytis holoxanthus** De Bach	
II. Coccidae		
7. *Saissetia oleae* (Mediterranean black scale)	*Coccophagus pulchellus* West, *Metaphycus flavus* Roward, *Metaphycus helvolus** Mercet, *Metaphycus lounsburyi** Howard	*Chrysopa* sp., *C. bipustulatus* L., *Exochomus flavipes* Goeze, *E. quadripustuatus* L., *Scutelista cyanea* Motsch. *Eublemma scitula* Ramb.
8. *Coccus hesperidium* (brown soft scale)	*Coccophagus scutarallis* Dalm., *Encyrtus lecaniorum* Mayr., *Metaphycus flavus* Howard	
9. *Coccus pseudomagnoliarum* (citricola scale)	*Coccophagus lycimnea* Walker	
10. *Ceroplates rusci* L. (fig wax scale)	*Coccophagus lycimnea* Walker, *Peraceraptocerus italicus* Masi, *Tetrastichus ceroplastae* Girault	*Eublemma scitula* Ramb., *Scutelista cyanea*
11. *Ceroplastes floridensis* (Florida white scale)	*Coccophagus lycimnia* Walker, *Tetrastichus ceroplastae* Girault	*Scutelista cyanea, E. quadripustulatus*
III. Margarodidae		
12. *Icerya purchasi* (cottony cushion scale)		*Rodolia cordinalis** Mulstant, *C. bipustulatus*

* Principal predator, parasite

continuous use of these chemicals increases the toxic residues, promotes resistance of the concerned scale insects, and enhances infestation of the mites. Zhang and Zhang (2000) attempted to control the infestation of *Icerya purchasi* and *Chrysomphalus aonidum* by spraying 200–250 times solution of mechanical oil emulsion + 800–1000 times solution of 40% Supracide in mid-May in China. They observed the effectiveness of this application for 25 days. It also controls red and rust mites.

Predators, Parasites and Biological Control

Citrus is one of the priority crops in which biological control of pests has proven most successful for effective control of several pests in all *Citrus*-growing countries. Predators and parasites have been duly acknowledged since 1868 when the *Citrus* industry in California was devastated by cottony cushion scale imported from Australia and was subsequently controlled by importing its predator, *Rodolia cardinalis* (Malsant) from Australia, multiplying and releasing the predators in the infested orchards in California. After this spectacular success, investigations on predators and parasites gathered momentum and natural occurrence of them was recorded in orchards of most of the citrus-growing countries. Intercountry importation of them began to investigate their effectiveness to manage scales and other pests of citrus. Hart (1978) conducted studies on the comparative effectiveness of different entomophagous insects to control scales in the USA and observed the differences (Table 8.3).

Table 8.3: Comparative effectiveness of different entomophagus insects to control scales in USA

Scale insect	Place of infestation	Principal enemy	Place of origin	Extent of control
California red scale	California	*Aphytis lignanensis* Comp	Southern China, India	Substantial
		Aphytis melinus DeB	Pakistan	Substantial
Florida red scale	Texas	*Aphytis holoxanthum* DeB	California	Complete
Purple scale	California	*Aphytis lepidosaphes* Comp.	China, Formosa, California	Partial

Natural enemies of scale insects exhibit wide biodiversity. Their effectiveness in controlling infestation depends primarily upon the geographic and climatic conditions of the area concerned. For example, *Aphytis lignanensis* has been found to be a successful parasite of the

California red scale in California, Morocco, and Argentina but not in Australia. *Comperiella bifaciata* that commonly parasitizes yellow scale in California and Florida, fails to do so in Australia. Parasites of Florida red scale (*Aphytis chrysomphali* and *Aphytis holoxanthus*) introduced into Israel, effectively controlled the scales in that country.

Demonstration of success of biological control in the laboratory and field, both led to its incorporation as the key option in integrated management. Riehl (1981) conducted field studies on integrated management by adopting both oil sprays and parasites. LV application of 93.5 L ha^{-1} of NR 415 oil in a volume of 935 L ha^{-1} of spray mixture to orchards in September under Californian conditions reduced the immature population of red scale and gave advantage to the parasite *A. melinus* in satisfactory control of it. LV application by a deposit of tiny discrete droplets avoids the flooding effect of dilute oil spray that destroys adult *A. melinus*. In Australia low range oil is sprayed where fruit is being produced for the fresh fruit market when scale density is high and parasite density low. Oil spray is hardly required in addition to parasites when fruit is grown for processing (Furness, 1981). Application of predators and parasites has been commercialized in Australia mainly to control red scale and snow scale. Commercial production of *A. lignanensis*, a parasite of red scale, and coccinellid beetle *Chilocorus carcumdantus*, a predator of the snow scale has been started in insectaries. These are routinely released in orchards at the rate of 10,000 per hectare in opportune time (Papacek and Smith, 1992). Application of biological enemies has been able to completely control the red scale in South Africa (Bedford and Grobler, 1981). Combination of oil and parasites is used in Italy to control black scale, fig wax scale, chaff scale and California red scale (Barbagallo, 1981). In southern California all the scales are controlled mostly by biological methods.

Surveillance and Monitoring

Success of integrated management depends on various biotic and abiotic factors. A perfect understanding is necessary of the phenology of the trees, biology of both the pests and their enemies under the concerned agroclimatic conditions, and the seasonal changes in their population. Surveillance and monitoring thus become very important in applying the options of integrated management, in particular the applications of enemies and chemicals. Takagi (1981) used a sticky suction trap to monitor parasitoids and pests. This trap consists of a fan (8 cm diameter) enclosed in a plastic box with a square opening (7 × 7 cm) at one end. It is suspended in a square frame on which a glass plate (20 × 20 cm) coated with an adhesive on its undersurface is placed. The distance between the trap opening and the sticky plate is 10 cm. The device is placed within a tree canopy with the fan side opening 10 cm above the leaf layer. The fan is propelled with an ac current. This trap samples the aerial population of flying insects and is equally effective day and night. The fan impels air onto the sticky plate at a constant rate of 72 m^3 per hour. It is possible to monitor several pests and parasitoids, such as soft scale, arrowhead scale, mealybugs, bark borers, and flower bud midge to schedule the time of application of parasitoids or chemical methods.

Monitoring, however, is usually done visually. Barbagallo et al. (1992) proposed an economic threshold and sampling methods for integrated management of citrus pests including a few scales. The intervention threshold, sampling method, and monitoring period for soft and armored scales are 1 female/10 cm twig or 3–5 nymphs/leaf, visual examination of 4 twigs (10 cm long/plant), and also 20 fruits/plant in case of armored scales respectively. Monitoring of soft scale is done in summer and winter and that for armored scale every month from summer to winter.

Grafton-Cardwell et al. (2000) attempted to monitor the population of yellow scale by using pheromones. They evaluated the effects of dose and field longevity of lures treated with synthetic female yellow scale sex pheromone ((5E)-6-isopropyl-3,9-dimethyl-5,8-decadienyl acetate) for monitoring flight activity of male yellow scale and found that a low dose (1–5 μ $lure^{-1}$) effective in monitoring the phenolgy and population densities of this insect.

Understanding the phenology of the trees, biology of the scales, and relationship between the pests and predators and parasites, as well as an appropriate monitoring system, it is now possible to properly manage scales by using several options, keeping biological control as the key one in the integrated system. This system is now gradually being upgraded by improving the quality of the predators and parasites, balancing the predators, and parasites, and pests, improving the techniques for multiplying the number, and incorporating several cultural practices and biotechnological methods to ensure situation more conducive to proper pest-parasite interactions.

Mealybugs

Mealybugs belong to the family Pseudococcidae. Seven species of these insects generally infest various species of *Citrus,* namely, (i) *Planococcus citri* (Risso) or *Citrus* mealybug, (ii) *Pseudococcus adonidum* L. and (iii) *Pseudococcus longispinus* (Targioni-Tozzetti) or long-tailed mealybugs, (iv) *Pseudococcus citriculus* Green or citriculus mealybug, (v) *Pseudococcus comstocki* Kuw., (vi) *Pseudococcus calceolarae* (Maskell), and (vii) *Pseudococcus affinis* (Mask.). *Citrus* mealybug is cosmopolitan,

C. comstocki originated in China and Japan, while other other species originated in different tropical and subtropical countries.

General Feature and Biology

Mealybugs normally feed on twigs. They may also infest leaves and fruits. On eclosion, nymphs mostly crawl to young twigs where they settle for feeding with very rare movement or locomotion.

Before full maturity, the dorsum of the body usually becomes covered with a dusting of powdery secretion. In some species, various areas of the dorsum may be bare, while in others dorsal wax may be arranged in lumps or irregular masses. Waxes are usually white and soft but in a few cases, the insects may be enveloped in a hard black case.

At full maturity, a large number of species secrete ovisacs in which eggs are deposited. These sacs may lie entirely beneath the body, forming a pad, while in others they may be longer than the insects themselves. Malleshaiah et al. (2000) studied the biology of *Planococcus citri* under laboratory conditions (temperature 25–29° C, relative humidity 65–70%) rearing them on pumpkin fruits. They observed that incubation lasted 3.35 days. The respective nymphal stages for the males and females were four and three instars. Nymphal development of males and females was completed in 20.05 and 28.10 days respectively. Fecundity ranged from 152–356 eggs.

The incidence of mealybugs is seasonal. Plant to plant dispersal of them is accomplished either by the first instar nymph being blown by the wind or by shifting of the developed nymphs and adult females by agents such as ants.

Franco (1992) studied the population dynamics of various mealybugs on a physiological time scale in Portugal. He considered three periods of citrus phenology: (1) the spring flush that includes shoot growth and blooming; (2) fruit set and development including cell division period, cell elongation stage, and maturation period; and (3) dormancy period. He conducted his studies on sweet orange. It was observed that the population dynamics of mealy bugs is directly influenced by the succession of the aforesaid phenological periods and their interactions with major climatic factors (temperature, humidity) and biotic factors (e.g. natural enemies). During periods 1 and 3, mealybugs are mainly dispersed in the tree canopy while in period 2, they show an aggregative tendency parallel with a progressive concentration of individuals on fruits and a rapid increment in mealybug populations is observed during this period.

From the beginning of period 1, overwintering individuals tend to migrate and settle at the bottom of young shoots. Development of these individuals seems to parallel shoot growth. They reach the adult stage mainly during the blossom period. Then at least some adult females migrate to the trunks and branches where they settle in bark crevices for oviposition.

Predators and Parasites

Furness (1981) recorded several natural enemies of long-tailed mealybug (*Pseudococcus longispinus*). Some are hymenopteras, while others are coccinellids and one a neuroptera. The hymenopteran insects are *Anagyrus fusciventris* (Girault) and *Moranila* sp., coccinellids *Rhizobius ruficollis* (Lea) and *Scymnus* sp., and neuroptera *Chrysopa* sp. Brun (1992) recorded three natural enemies of citrus mealybug (*Planococcus citri*): *Cryptolaemus montrouzieri* Muls, *Leptomastix dactylopii* How., and *Nephus reunioni* F. Spicarelli et al. (1992) demonstrated the biological control of *Citrus* mealybugs by releasing artificially cultured *Leptomastix dactylopii* in Italy. They observed variation in host incidence and parasitism in different seasons. They also recorded the natural incidence of the predators *Leucopsis* spp., *Scymnus includens* Kirsch, *C. montrouzieri*, and chrisopid larvae preying upon *P. citri* in a particular season.

Monitoring and Management

Integrated methods keeping the application of natural enemies as the key option are generally used in management of mealybugs. *Cryptolaemus montrouzieri* is used in Italy in control of this pest by biological method (Barbagallo, 1981). *Anarhopus sedneyensis* Timb and *Hungariella peregrina* (Compere) are used to control long-tailed mealybugs (*Pseudococcus longispinus*), while *Leptomastix abnormalis* (Girault) and *Cryptolaemus montrouzieri* are used to control *Citrus* mealybugs.

Monitoring is the most important component of integrated management. Barbagallo et al. (1992) suggested that visual examination of 10 fruits plant^{-1} is to be done every 20 days from summer to autumn in Italy. Intervention is to be done when 5–10% fruits become infested by *Planoccocus citri*.

The modern method for monitoring mealybugs employs pheromone traps. Moreno and Kennett (1981) observed that male mealybugs could be trapped by pheromone producing virgin females caged in 0.5 L cartons. Males were lured from least 34 meters away from the source of infestation. These authors further observed that tagged males could fly to pheromone traps 183 meters from a central point of release, i.e. the flight range of *P. citri* during the warmer part of the year. They analyzed this pheromone as 2,6-dimethyl-1-5-hepta-dien-3-oacetate. In the same year Bierl-Leonhardt et al. isolated and synthesized the sex pheromone of *P. citri* into (IR-cis)-(+)-2,2-dimethyl-3-(1-methyl-ethyl)cyclo butanemethanol acetate. Since this compound can be synthesized, it is now widely used for both monitoring and control of *P. citri.*

Raciti et al. (2001) in Sicily used yellow pheromone traps to monitor *Planococcus citri* and the indigenous and introduced parasitoids (*Leptomastix dactylopii*: introduced, *Anagyrus pseudococci*: indigenous). They found that introduced *L. dactylopii* populations were a decisive factor in control of this mealybug, with parasitism reaching as high as 70%. Since *L. dactylopii* cannot survive in cold temperature, however, it has to be introduced annually for control of *P. citri.*

Removal of Mealy Bugs from Fruits

Gould and McGuire (2000) evaluated hot-water treatment and insecticide-coatings to disinfect limes of mealybugs *Planococcus citri* and *Pseudococcus odermatti*. They found that a 20-minute 49° C hot-water immersion treatment to be effective in killing mealybugs and all other arthropod pests. They treated 7,200 limes with 1,308 insects and found no survivor after the treatment. Ampol (petroleum-based oil) coating gave up to 94% mortality of mealybugs, which is not sufficient to provide quarantine security. It is effective as a postharvest dip before shipment.

TRUNK BORER (*ANOPLOPHORA VERSTEEGI* (RITS.))

Symptoms

Severe infestation of trunk borer may produce dieback-like symptoms.

Biology

The trunk borer or larvae of *Anoplophora* beetle bore the trunk of plants. Eggs are laid during May–June in subtropical climate within the barks on the trunk of the tree in a slit dug by the mandibles of the adult females. Oviposition takes place after dusk. The early larval stages are usually spent underneath the bark where they live by forming galleries in the sapwood. Excreta come out through the entry hole of the larvae as an exudation and chewed wood. The area of initial damage appears as a slight swelling on the trunk. Later, the larvae dig into the heartwood and gradually reach the central portion of the trunk from where they move upward by gnawing a vertical tunnel. The entry points of the larvae into the heartwood appear as irregular pits. These points can be easily seen after removal of the bark of the infested plant.

Mature larvae gradually transform into pupae within the tunnel made in the heartwood of the infested plant. The pupa gradually transforms into an adult that emerges from the trunk. The exit hole of the adult is bigger and circular. These holes are prepared by the larvae before pupation within the tunnel of the heartwood of the trunk.

LEAF MINERS

The leaf miner (*Phyllocnistis citrella* Stainton: Phyllocnistidae: Lepidoptera) remained a minor pest until the 1980s, after which it became a very serious pest, particularly in nurseries in all *Citrus*-growing countries of the world. Another Lepidopteran insect, *Citrus* flower moth (*Prays citri*), has quarantine significance but it has not yet assumed the status of a serious pest. For example, there was no leaf miner in Spain but as soon as it was introduced in 1993, it gained the status of a very serious pest of nursery trees there. The same is true of several other countries.

Symptoms and Feeding Behavior

Leaf miners cause serious damage to young leaves and stems. Infested leaves curl variously and later appear crumpled (Plate IV: Figure VIII.1).

Apodous larvae of the leaf miner attack only the younger and tender leaves and make serpentine mines in them. The larvae feed on the epidermal cells of leaves, leaving the remaining smooth waxy cuticle that protects the larva. Larvae feed in zigzag lines and severely infested leaves become misshapen. Infestation and severity of the pest depend on several factors. In a general way, leaves have certain compensatory abilities in photosynthesis and tissue repair. But under favorable conditions, severe infestation may occur depending on the host, climate, agrometeorological, and several other ecological conditions. There is always a threshold level for severe infestation. Wang et al. (1999) in China developed a multifactor economic threshold model on the basis of several ecological factors, varieties, hydrothermal conditions in the citrus gardens, yield, and natural enemies. They could set the economic threshold in Xuesan citrus as 12.21 larvae of the first instar per tender leaf.

Boughdad et al. (1999) observed in Morocco that each larva destroys 4 to 100% of the leaf surface before reaching the pupal stage. They also found a correlation between the mined surface and the total surface of the leaf on the one hand, and between the destroyed surface and the severity on the other.

Biology and Ecology

One female insect may lay 12–77 eggs on the undersurface of tender leaves and occasionally on the upper surface that undergo incubation for of 2–10 days. The larva passes through four instars. The larval period ranges from 5–20 days depending on the climate and season. The fully-grown larva curls a small part of the edge of the mined leaf over its body, spins a white cocoon, and pupates inside. The pupal period varies from 6 to 22 days. In mining the epidermal layer of the leaves, the larvae make a series of galleries. On removing the cuticle above such galleries, the flattish small creamy yellow larvae can be seen. When full grown, the larvae turn crimson colored. After hatching the larvae enter the laminar tissue and mine through the epidermis, feeding on the tissues and leaving the thin translucent cuticle above. In very young leaves the larvae may leave black stringy material behind them as they advance by feeding. This trail is actually their excreta. The adult moth is shiny, silvery white, its forewings with brown stripes and a prominent brown spot near the apical margin, and hindwings white. Both wings are fringed with minute hairs.

Boughdad et al. (1999) in their study on the biology of the leaf miner in Morocco observed that the insect lives 13 to 45 days; 86% of the eggs laid hatched; but 42 to 86% of the hatchlings are normally killed by abiotic and biotic factors before reaching the adult stage. This insect may exhibit up to 8 generations per year.

Ujiye (2000) in Japan found that the adult leaf miner overwinters within the canopy of citrus trees in the warmer districts and that overwintering females begin to oviposit in mid-March. The moth passes through 9–10 and 5 generations per year in the southwestern and northwestern parts of the citrus belt respectively. Developmental zero and effective heat-sum for development from egg to adult emergence, they found as 12.1° C and 206 day-degrees. 80% or more larvae are killed due to unknown factors before pupation during summer-autumn seasons. It may primarily be due to parasitoids. The dominant killer parasitoid in the main-land was found to be *Sympiesis striatipes* and those in the southern islands were *Cirrospilus ingenuus* and *Citristichus phyllocnistoides*. Interestingly, a sex attractant (7Z, 11Z)-7, 11-hexadecadienal shows high activity for indigenous male moths but appears ineffective for those introduced from foreign countries.

Biological Control and Integrated Management

Several pesticides can effectively control the infestation of leaf miners. Mansanet et al. (1999) in Spain applied confidor 200 SL (imidacloprid) in foliar treatment. Application can also be made via the irrigation system or the undiluted chemical can be applied to the bark. Cruz and Dale (1999) in Peru also applied imidacloprid (confidor R 350 SC) by spray or drenching technique and found < 80% efficacy for a period of 8 weeks by drenching and 10–14 days by spraying. BeiXueFen et al. (1999) in China found that spraying 1,000 or 2,000 x solution of phosmet provided 98.9 and 95.3% mortality of leaf miners. When 2,000 x solution of carbosulfan was applied the mortality was 94.1%. BeiXuFang et al. (1999) in the same country used 10% imidacloprid WP (1: 1,000) and obtained 98.9% mortality 7 days after application. They found the superiority of shoot protection effectiveness value of this application much higher than that obtained with 20% carbosulfan (1:2,000). LaiBang Hai et al. (2000) observed that abamectin-petroleum (24.5%) 1:500 controlled 95.8% of infestation within 7 days of application. Zhang et al. (2001) concluded that the best insecticide for controlling leaf miner is a 100-fold solution of 3% acetamiprid, which killed 97.8% miners 3 days after application. Borad et al. (2001) found 10% (w/v) spray of neem and Naffatia (*Ipomoea fistulosa* (*I. carnea*)) effective in managing the miner population in nurseries.

Biocontrol

Abbassi et al. (1999) did a comparative study on the efficacy of the indigenous (*Cirrospilus strictus, Pnigalio* sp.) and introduced (*Ageniaspis citricola* and *Semielacher petiolatus*) biological enemies to control leaf miner on grapefruit in Morocco. They observed that introduced *S. petiolatus* was more adaptive in summer conditions with high temperature and low humidity and showed 66–67% parasitization.

Alkhateeb et al. (1999) in Syria isolated and identified a few parasitoids. These are *Ratzeburgiola incompleta, Cirrospilus* sp. nr. lyncus, *Cirrospilus ingenuus, Semielachar petiolatus* and *Neochrysocharis* sp. Conti et al. (2001) described indigenous occurrence of *Cirrospilus pictas, C. vittatus, Diglyphus isaea, Pnigalio agraules, Asecodes delucchii, Chrysocharis pentheus, Neochrysocharis formosa, Apotetrastichus postmarginalis, A. sericothorax* and *Aprostocetus* sp. in Italy and noted the effect of rain and temperature on their distribution. Legaspi et al. (2001) found that effectiveness of parasitoids differed with the difference in geographic location. The overall percentage parasitism by *Zagrammosoma multilineatum* was higher in Mexico than in Texas, USA. Amalin et al. (2001) found hunting spiders to be efficient predators of leaf miners. Comparative predation efficiency of different spiders was as follows: *Hibana velox* > *Cheiracanthium inclusum* > *Trachelas volutus.*

Amalin et al. (2001a) studied the prey capture behavior of these noctuid spiders. *H. velox* detects its prey by sensing vibrations of the substrate induced by the concealed prey. Movement of *P. citrella* larvae and prepupae appears to create vibrations of leaf substrate serving as cues for the spiders to locate them. The spider punctures the mine, immobilizes the larva, then bites and sucks the larval body fluid. In another behavioral pattern, the spider makes a slit in the mine, uses its forelegs to pull the larva or prepupa out, of the mine, holds the prey securely and bites it, then regurgitates digestive juices into the prey and ingests the predigested liquid tissue. The spiders start feeding at the second instar stage. Consumption increases as they develop to later instars, with maximum consumption in the fourth instar. But *H. velox* fails to complete its life cycle with miners as the only food whereas *C. inclusum* and *T. volutus* do.

Cultural Control

Ateyyat and Mustafa (2001) found several cultural practices to minimize miner infestation in Eureka lemon. They observed an increase in number of larvae, pupae, and their parasitization as well, with an increase in use of urea. The number of pupae peaked when urea was applied 1 kg tree^{-1}. They recommend that growers subsequently should avoid applying > 85 g urea tree^{-1}. Summer pruning in which the pruned branches are placed under the trees decreases the number of live larvae and pupae without affecting their parasitization.

BUTTERFLY

There are two types of butterflies that occasionally infest citrus, notably lemons. These are Orange dog butterfly (*Pappilio cresphontis* Cram) and Lemon butterfly (*P. demoleus*). They feed on leaves and lay eggs on the under surface of the tender leaves.

Biology

The butterflies look beautiful; the wings bear black and yellow markings and in the hind-wings toward the inner margin a brick-red oval patch occurs. A female butterfly lays 75–120 shiny grayishyellow eggs in 2–5 days. Incubation is 3–8 days. There are 5 larval instars. The larval stage lasts for 11–40 days, after which the larva pupates in a naked chrysalis attached to the plant by two fine strands of silk in the form of a girdle. Adults emerge in one week in summer but in 10–20 weeks in winter. The life cycle is normally completed in 18–40 days; depending on temperature, it may extend up to 145 days.

These flies have several alternate hosts, e.g. *Murraya koeningii, Aegle marmeloes, Psoralia cordifolia, Feronia elephantum*, etc.

Natural Enemies

A yellow wasp, *Polistes herebreus,* and the praying mantis, *Crebrator gemmatus* usually attack butterfly larvae. There are also several parasitoids. Some, *Apanteles papilionis* Viereck of Braconidae, *Melalopacharops* sp. of Ichneumonidae, *Pteromalus puparium* (L.) of Chalcidae, *Holcojoppa coelopyga* of Ichneumonidae, are pupal parasitoids.

FRUIT FLIES

Three types of fruit flies are of quarantine significance: the Mediterranean (*Ceratitis capitata* Weid), South American fruit fly (*Anastrepha fraterculus* Weld) and Oriental fruit fly (*Dacus dorsalis* Hend/*D. caudatus* (Fabr)). There is also another fruit fly called the Caribbean fruit fly (*Anastepha suspense* Loew). Other species of *Dacus* also infest *Citrus* fruits, in particular *D. oleae* (Gmelin), *D. tryoni* (Froggett). Another species of *Anastepha* called *A. ludens* (Loew) is also found to infest.

Symptom

The only visible symptom of fruit fly infestation is fruit drops. When the maggots feed on the fruit pulp brown rotten patches may appear on the attacked fruits.

Biology

Fruit flies are small to medium size insects, usually with spots or bands on the wings. Wing venation is characteristic in that the subcosta bands are apically forward almost at a right angle and then fade.

Adults oviposit in mature fruits and the larvae feed and develop in the fruit pulp and are not externally visible. On splitting the infested fruit, numerous creamy white larvae are readily visible in the pulp of the fruit.

Monitoring and Control by Baiting

Katsoyannos et al. (1999) did a comparative study on the trapping of Mediterranean fruit flies (Med-fly) using various traps and baits in Greece. They used two types of traps, the International Pheromone Mc Phail Trap (IPMT) and the Tephri Trap. Two types of baits were used, synthetic female targeted lures and a standard protein bait (Nu-Lure) and borax. The synthetic Lure contained ammonium acetate, 1,4 diaminobutane (putrescine) and trimethylamine. They found wet IPMT baited with synthetic attractants the most attractive method; two times more female and 1.8 times more total Med-flies were caught than obtained with traps baited with Nu-Lure and borax. But dry traps were more effective for females and more practical for mass trapping and monitoring than traps baited with protein solutions. Cornelius et al. (2000) in Hawaii, USA did a study on the effectiveness of different baits. They found protein-fed (Nu-Lure) females (10–12 days old) were more attracted to fruit odors than to protein-odors. But mated deprived females (10–12 days old) and unmated protein-fed females (2–3 days old) were equally attracted to fruit and protein odors. They further evaluated the attractiveness of the commercially available protein baits, Nu-Lure, Trece A.M. Spercharger, and Bio-Lure. According to them, Nu-Lure was more effective for capturing females. Dias and Arthur (2000) in Brazil used Valencia as traps and compared the effectiveness of variuos attractants, viz. vegamine and ammonium sulfate (7:1), sugarcane syrup and ammonium sulfate (3:1), vegamine 10% and concentrated maracock juice (25%). The effectiveness was tested for a period of one year. During this period they caught 1,004 fruit flies, of which 72.91% were *C. papitata* and 27.09% comprised of of 8 species of *Anastrepha*; with *A. fraterculus* predominant. Maracock juice was proven an efficient attractant. The percentages of catch with this were 43.34% *C. capitata* and 46.10% *Anatrepha* sp.

Miranda et al. (2001) did a field evaluation of female-targeted trapping systems for Med-fly in Spain. They baited both IPMT and Tephri traps with a three-component food-based synthetic attractant called 3FA. This attractant was composed of putrescine, ammonium acetate, and trimethylamine. They also used IPMT-baited with Nu-Lure and Delta traps baited with Trimed-Lure. They observed a high population level during autumn/winter. They found 3 FA very effective in catching females and also at a low population level, particularly in Tephri traps both in wet and dry condition. But Trimed-Lure is very useful in detecting Med-flies in the field at a very early stage.

Nishida et al. (2000) worked on a new attractant, Alfa-copaene for male Med-fly. It occurs as a minor component in the essential oils of various plant species including its hosts, such as orange, guava, and mango. Using a plastic leaf model treated with (+) Alpha-copaene, they observed that it affects virgin females, provoking pseudomale courtship behavior. Exclusive mating on artificial leaves suggests the potentiality of the compound for its use in the Med-fly management.

Considering its quarantine significance, Hallman and Rene-Martinez (2001) used gamma irradiation to treat Rio Red grapefruit against Mexican fruit fly to prevent adult emergence from third instars. Minimum absorbed dose was 58 or 69 Gy, commercially three times the minimum required dose. They found no change in the soluble solid contents, titratable acidity, appearance, and organoleptic quality of the treated fruits up to the dose of 500 Gy. This treatment is equally applicable to orange, tangerine, mandarin, etc.

MITES

Mites are very small microanimals. Several species are found on various citrus species either as pests or predators. These are strange components of the citrus canopy ecosystem keeping the biological balance. They are mostly found with scales and other Arthropods and become aggressive during incompatible use of pesticides. Much is yet to be studied on their incidence and the interrelations between them and various insects.

Mites are broadly divided into three families: Eriophyidae, Tetranychidae and Phytoptipalpidae. Common mite pests found within these families are given in Table 8.4.

Table 8.4: Common mite-pests of citrus

Family	Scientific name	Common name
Eriophyidae	*Aceria sheldoni* Ewing	Citrus bud mite
	Phyllocoptruta oleivora Ashm.	Citrus rust mite
	Aculus pelekassi Keifer	Citrus rust mite
Tetranychidae	*Eotetranychus annecki* Meyer	Texas citrus mite
	Panonychus citri McG.	Citrus red mite
	Tetranychus urticae	Two-spotted spider mite
	Eotetranychus sexmaculatus Riley.	Six-spotted mite
Phytoptipalpidae	*Brevipalus californicus* Bks.	Citrus flat mite/false spider mite
	Brevipalus australis Tucker	Citrus flat mite/false spider mite
	Brevipalus phoenicis Geijskes	Broad mite/reddish black flat mite
	Brevipalus lewisi McG.	Citrus flat mite
	Brevipalus obovatus Donnadieu	

In addition to the mites in the Table, there are two others of quarantine significance: Citrus brown mite (*Eotetranychus orientalis*) and Citrus silver mite (*Polyphagotersonemus latus*).

CITRUS BUD MITE (*ACERIA SHELDONI* EWING)

Symptoms and Biology

The bud mite infests flowers and buds. Infestation of this mite causes malformed twigs, leaves, and fruits. Completion of its life cycle requires 10 days to one month depending upon climatic conditions. This mite infestation is normally restricted to lemons.

Moderate temperature, humidity, new growth and deficiency of potassium favor development of the bud mite. The threshold temperature for embryonic development of this mite is 9° C; most successful eclosion takes place at 25° C and 98% relative humidity. High temperature (> 34° C) and low humidity (35–40%) normally reduce the population.

Natural Enemies and Control

Agistemus exsertus Gonzales et Stigmaeid mites are predators of bud mites and in nature, generally keep this pest under control. In case of severe infestation, mineral white oil spray can control this pest. Application of broad-spectrum pesticides, particularly those with trophobiosis or those that adversely affect predatory mites, is not recommended (Barbagallo et al., 1992).

Citrus Rust mite (*Phyllocoptruta oleivora* Ashm./ *Aculops pelekassi* (K.)/ *Tegolophus australis* Kiefer)
Citrus rust mite is more or less cosmopolitan in distribution. This mite normally affects fruits.

Biology

In humid subtropical warm climate, the population of these mites usually increases at a very fast rate, developing densities of 70–100 mites cm^{-2} on fruits surface within 7–10 days depending on temperature. Limiting temperature for development of this mite is 17.6 and 31.4° C; the optimum temperature for development is 24.5° C.

Infestation takes place by the spermatophores. They are deposited on the plant surface by male species. The shape of the spermatophores varies with the species, and production per day depends on the season; the usual production rate is 16 per day. During their lifetime production may reach 145–155 spermatophores. Virgin females take up the sperm masses produced by the spermatophores. These mites overwinter as adult females, which remain hidden between bud scales.

Natural Enemies and Control

Phytoseiid mites are normal predators of rust mites. Some of these mites are *Amblyseius deleomi* Muma et Denmark, *A. lentiginosus* Denmark and Schicha, *A. elinaeschida*, Coccinellid beetles such as *Halmus* (*Orcus*) *chalybeus* Boisduvan, *Serangium bicolor* Blackburn, and *Stethorus nigripes* Kapur (Beatle, 1978). *Hirsutella thompsonii* is an efficient fungal predator of these mites.

Selective acaricide chlorobenzilate at the rate of 0.038 kg L^{-1} when applied by ground equipment at high volume (8 L plant^{-1}) using a pistol-type sprayer under pressure of 17.85 kgf cm^{-2} can control citrus rust mites. The threshold level of infestation for such application is 1% fruits with 75 mites cm^{-2} (Trevizoli and Gravena, 1984).

Dybas et al. (1984) reported the use of a novel chemical "Avermectin" B 1 (MK 936), a macrocyclic lactone isolated from *Streptomyces avermilitis*. It is toxic at a dose less than 1.0 ppm (5.0–4.0 g a.i. ha^{-1}) with low residual toxicity to beneficial insect complexes. It has been reported by several workers that application of organophosphates and many other miticides develop either tolerance or resistance in rust mites (Papacek and Smith, 1992). Lu and Lu (2000) applied 0.15% enhanced abamectin and found quick action that persisted for 20 days; morality was 97.2–99.9%. The effective concentration was 1,500–2,500 X. They also observed the effectiveness of 20% dicofol at a concentration of 800 X. Ying et al. (2000) on the other hand, found Dithane M 45 (mancozeb) at 600 X, solution the most effective with a residual effectiveness for 34–60 days.

Spider Mites

Spider mites are of several types: red mite, Texas citrus mite, carmine spider mite, six-spotted mite, two-spotted mite, etc.

Red Mite (*Panonychus citri* McG.)

Symptoms

This mite normally feeds on leaves. Infested leaves show stippling and bronzing. Heavily infested leaves may drop following twig dieback.

Biology

The life cycle of the red mite consists of eggs laid on the surface of leaves, six-legged larvae at the immobile stage, eight-legged ones at the mobile stage, protonymph, deutonymph, and adult males and females. Under warm conditions, this mite can go from egg to adult within 12 days (Jones and Morse, 1984). The maximum and minimum temperatures for the development of this mite are 30° and 19° C respectively. It may complete 10–12 generations per year. These mites repeatedly probe and retract in a single feeding puncture, allowing plant fluids to flow onto the surface.

Monitoring, Natural Enemies and Control

Monitoring of citrus red mite populations is accomplished by watching for the appearance of adult females on tender full-size leaves and for stippled leaf injury. The economic threshold for applying control measures is tentatively considered 2 adult females per leaf (Pehrson et al., 1984)

Furuhashi et al. (1981) constructed a computer model to simulate the population fluctuation of citrus red mite and to forecast the occurrence of population density. In this model seven developmental stages of the mite (egg, larva, protonymph, deutonymph, preovipositing female, ovipositing female, and nonovipositing female) and temperature, rainfall, and typhoon as environmental factors, impact of pesticide spraying and natural enemies on the dynamics of the pest were taken into account. The simultaneous ordinary differential equations were solved by the "Runge-Kutta" method. The time unit used was one day. The model was programed in FORTAN.

The simulated results were comparable to field data. The model depended on environmental factors.

Integrated management: Ming-Dau et al. (1981) and Gravena et al. (1992) demonstrated the integrated management of citrus red mite. Effective biological control can be achieved by the polyphagous phytoseiid mite *Amblyseius newsami* Evans, when weeds *Ageratum conyzoides* Linn. and *Eupatorium pauciflorum* are grown in the orchards. They observed that the weeds reduced summer temperature 40–45° C to < 35° C, increased humidity, suppressed growth of other weeds, and provided food for predaceous mites through pollens. These weeds can also be used as green manure.

Kim et al. (2000) recorded several natural enemies in Korea: *Oligota kashmirica benefica, Chrysopa pallens, Propylea japonica, O. yasumatsui, Stethorus punctillum, Orius sauteri, Scolothrips takahashii, Amblyseius womersleyi,* and one unidentified species. The effectiveness of management of the mite has yet to be worked out.

Li-HongYun et al. (2001) studied the resistance of various *Citrus* species to red mite infestation. They found *Atalantia buxifolia* (*Severinia buxifolia*) to be very resistant followed by an American strain of *Poncirus trifoliata, Citrus limonia,* and *P. trifoliata* cultivar Luxizhi.

Chemical control: Several miticides are available to control red mite. These include various oil formulations, dicofol (Kelthane), propargite (Omite), cyhexatin (Plictran), oxythioquinox (Morestan), fenbutatin-oxide (Vendex), amitraz (Baam or Mitac), and avermectin (Abamectin). But many of these chemicals may cause appreciable leaf burn on young, tender foliage, especially when the temperature exceeds 32° C (Morse, 1986). Avermectin B 1 has also been reported to control the red mite (Dybas et al., 1984).

Lei Hui De et al. (2000) found increasing fungicide resistance in red mite. They tested the effectiveness of Fumanlang of unknown composition manufactured by Nanjing Agrochemicals in China to be more effective than NC 129 (pyridaben). The effect of 20% solution of this compound at 1,500, 2,000 and 2,500 X concentration persisted for 20 days or more.

Two-spotted Spider Mite (*Eotetetranychus urticae*)

The two-spotted spider mite is a major pest of a wide variety of agricultural crops. It causes substantial trouble on citrus in greenhouse situations as it can withstand high temperature.

The biology, monitoring, biological and chemical control methods are more or less similar to those described for red spider mite.

Six-spotted Mite (*Eotetranychus sexmaculatus*)

Six-spotted mite is a relatively uncommon pest in citrus nurseries. It is, however, a concern in nurseries located in the coastal area. Biology, monitoring, biological and chemical control methods of this mite are more or less the same as those for red spider mite.

Natural Enemies of Spider Mites

A large number of insects and mites have been recorded as natural enemies of spidertetranychid mites (Jeppson, 1977; Trevizoli and Gravena, 1984), viz.

Conwentzia psociformis Curt. (*Neuroptera:* Coniopterygidae); *Stethorus punctilum* Weise (Coleoptera: Coccinellidae); *Oligota* sp., *O. pygmea* Solier (Staphylinidae); Larvae of Cecidomyidae and Syrphidae; *Typhlodromas floridanus* (Muma), *Phytoseiuslus persinilis* Athias-Henriot, *Typhodromus rickeri* Ghant, *Cydnodromus* (A.) *californicus* (McG.) *Euseius* (*A.*) *stipulatus* (Athius-Henriot), etc. (Phytoseiidae).

Brevipalus Mites/Leprosis Mites / False Spider Mite

In addition to the natural infestation causing damage to plants, several *Brevipalus* mites (*B. phoenicis, B. obovatus, B. californicus*) also transmit one or more Rhabdovirus or Rhabdovirus-like diseases of *Citrus* (leprosis), coffee (ringspot), *Lingustrum lucidum* (ringspot), passionfruit (green spot), and orchids (fleck) (Chagas et al., 2001). In the absence of viruses, they themselves may cause serious damage to plants.

Symptoms

Brevipalus mites produce several types of scars on twigs, leaves, and fruits. These scars resemble those of leprosis. In grapefruit, irregular brownish blemishes vary in size from 1 to 30 mm or more. There may also be proliferation along the main stem. High population may cause diffuse chlorotic spotting called "phoenicis blotch". These mites suck the plant juice into the pharynx through grooves formed by operating stylets. While feeding, they may release citrus oil and inject toxin into the plant that may be local or systemic. Some of them may also transmit the "leprosis" virus.

Important species are: *B. californicus* (Banks), *B. phoenicis* (Geijskes), *B. lewisi* McG. and *B. obovatus* Donnadieu.

Rezk (2001) studied the population dynamics of *B. obovatus* on Washington Navel in Egypt. The peak population appears in early spring and autumn at an average of 0.71 and between 0.52 and 0.51 mites^{-1} leaf respectively. When reared on citrus and mint 25° C and 65% (relative humidity), the mean incubation period is 3.07 and 4.66 days on citrus and mint respectively. Eclosion rate is 87.49 and 72.9% for both hosts. The active larvae and deutonymphal stages last on average of 2.76, 2.63 and 3.21 days on citrus and 3.13, 3.23, and 4.53 days on mint. Female longevity averaged 34.26 and 41.16 days in citrus and mint respectively.

REFERENCES

Abbassi M, Harchaoui L, Rizqi A, Nadori EB and Nia M (1999). Biological control against the citrus leaf miner *Phyllocnistis citrella* Stainton (Lepidoptera: Phyllocnistidae). *Proc. Fifth Int. Conf.* Pests in Agriculture, Part 3, Montpellier, France, pp. 609–615.

Alkhateeb N, Raie A, Gazal K, Shamseen F, and Kattab S (1999). A study on population dynamics of citrus leaf miner (*Phyllocnistis citrella* Stainton) and its parasitoids. *Arab. J. Plant Protection* 17: 60–65.

Amalin DM, Pena JE, and McSorley R (2001). Predation by hunting spiders on citrus leaf miner (*Phyllocnistis citrella* Stainton) (Lepidoptera: Phyllocnistidae). *J. Ento. Sci.* 36: 199–207.

Amalin DM, Reiskind J, Pena JE, and McSorley R (2001a).Predatory behavior of three species of sac spiders attacking citrus leaf miner. *J. Arach.* 29: 72–81.

Argov Y, Rossler Y, Voet H, and Rosen D (1999). Spatial dispersion and sampling of citrus whitefly. *Agricultural and Forest Entomology* 1 (4): 305–318.

Argyrou LC and Mourikis A (1981). Current status of citrus pests in Greece. *Proc. Int. Soc. Citriculture,* vol. 2, pp. 623–626.

Ateyyat MA and Mustafa TM (2001). Cultural control of citrus leaf miner *Phyllocnistis citrella* Stainton (Lepidoptera: Phyllocnistidae) on lemon in Jordan. *Int. J. Pest Management* 47: 285–288.

Aubert B (1987). *Trioza erytreae* Del Guercio and *Diaphorina citri* Kuwayama (Homoptera: Psyllidae), the two vectors of citrus greening disease: biological aspects and possible control strategies. *Fruits* 42: 149–162.

Aubert B and Xia YH (1990). Monitoring flight activity of *Diaphorina citri* on Citrus and Murraya Canopies. *In*: Rehabilitation of Citrus Industry in the Asia Pacific Region (eds. B Aubert, S Tontyaporn and D Buangsuwon). UNDP-FAO Regional Project RAS/86/022: 181–187.

Barbagallo S (1981). Integrated control of citrus pest in Italy. *Proc. Int. Soc. Citriculture,* vol. 2, pp. 620–623.

Barbagallo S, Rapisarda C, Siscaro G, and Longo S (1992). Status of the biological control against citrus whiteflies and scale insects in Italy. *Proc Int. Soc. Citriculture,* vol. 3, pp. 1216–1220.

Beatle GAG (1978). Biological control of citrus mites in New South Wales. *Proc. Int. Soc. Citriculture,* vol. 2, pp. 156–158.

Bedford ECG and Grobler H (1981). The current status of the Biological Control of the Red Scale *Aonidiella aurantii* (Mask) on citrus in South Africa. *Proc. Int. Soc. Citriculture,* vol. 2, pp. 616–619.

Bei XueFang, Jiang LiYing, Chi XueLiang, He-Qing Wei, Bei XF, Jiang LY, Chi XL, and He QW (1999). Experiment of spraying Imidan for control of citrus leaf miner. *South China Fruits* 28: 23.

Bei XuFen, Jiang LiYing, Chi XuLiang, He Qing Wei, Bei XF, Jiang LY, Chi XL, and He QW (1999). Control of citrus leaf miner with imidacloprid WP. *Plant Protection* 25: 52–53.

Benfatto D, Conti F, Frittitta C, Perrotta G, Raciti E, and Tumminelli R (2000). Spray trials results against a new citrus thrips *Pezothrips kellyanus* (Bagnall). *In*: GF 2000, Atti, Giornate Fitopatologiche Perugia, vol. 1, pp. 381–386.

Bierl-Leonhardt BA, Moreno DS, Schwarz M, Forster HS, and Plimmer JR (1981). Isolation, identification and synthesis of the sex pheromone of the citrus mealybug, *Planococcus citri* (Risso). *Tetrahedron Letters* 22: 389–392.

Borad PK, Patel MJ, Vaghela NM, Patel BH, Patel MG, and Patel JR (2001). Evaluation of some botanicals against citrus leaf miner (*Phyllocnistis citrella*) and psylla (*Diaphorina citri*). *Indian J. Agric. Sci.* 71: 177–179.

Boughdad A, Bouazzaoui Y, and Abdelkhalek L (1999). Pest status and the biology of the leaf miner *Phyllocnistis citrella* Stainton (Lepidoptera: Phyllocnistidae) in Morocco *Proc. Fifth Int. Conf.* Pests in Agriculture, Part 2, Montpellier, France, pp. 251–259.

Brun P (1992). Integrated pest control in Corsican citrus orchards. *Proc. Int. Soc. Citriculture,* vol. 3, pp. 968–970.

Cerameli M, Morales P, and Godoy F (2000). Presence of the Asiatic citrus psyllid *Diaphorina citri* Kuwayama (Hemiptera: Psyllidae) in Venezuela. *Boletin de Entomologia Venezolana* 15: 235–243.

Chagas CM, Rosesetti V, Colariccio A, Lovoisolo O, Kitajima EW, and Childers CC (2001). *Brevipalus* mites (Acari: Tenuipalpidae) as vectors of plant viruses. *Proc. 10th Int. Cong. Acarology* (eds. RB Halliday, DE Walter, HC Proctor, RA Norton, and MJ Collof), pp. 369–375.

Chen BingXu, Huang HanJie, Liu JingMei, Cai MingDuan, Chen BX, Huang HJ, Liu JM, and Cai MD (1999). Studies on the binomics and control of citrus spiny white fly. *Acta Phytophylacica Sinica* 26: 338–342.

Conti F, Raciti R, Campo G, Siscaro G, and Reina P (2001a). Biological control of *Phyllocnistis citrella. Informatore Agrario* 57: 57–59.

Conti F, Tumminelli R, Amico C, Fisicaro R, Raciti E, Frittitta C, Perrotta G, Marullo R, and Siscaro G (2001). The new thrips of citrus fruits *Pezothrips kellyanus. Informatore Agrario* 57: 43–46.

Cornelius ML, Nergel L, Duan JJ, and Messing RH (2000). Responses of female oriental fruit flies (Diptera: Tephritidae) to protein and fruit odors in field cage and open field condition. *Environ. Ent.* 29: 14–19.

Cruz R and Dale WE (1999). Control of the citrus leaf miner by drench treatment with imidacloprid on desert soils in Peru. Special Issue: Imidacloprid. *Pfanzenschutz Nachrichten Bayer* 52: 310–319.

Dias MCR and Arthur V (2000). Survey of the fruit flies in citrus orchards with four different attractive in Piracicaba-SP, Brazil *Revista de Agricultura Piracicaba* 75: 415–423.

Dybas RA, Unduraga JM, and McCoy CW (1984). Pests and their control. *Proc. Int. Soc. Citriculture,* vol. 2, pp. 449–451.

Franco JC (1992). Citrus phenology as basis to study the population dynamics of the citrus mealybug complex in Portugal. *Proc. Int. Soc. Citriculture,* vol. 3, pp. 929–930.

Furness GO (1981). The role of petroleum oil sprays in pest management programs on citrus in Australia. *Proc. Int. Soc. Citriculture,* vol. 2, pp. 607–611.

Furuhashi K, Nishino M, Muramatsu Y, and Shiyomi M (1981). Simulation model for the forecasting of occurrence of citrus red mite *Panonynchus citri* (McGregor) in citrus orchards. *Proc. Int. Soc. Citriculture,* vol. 2, pp. 654–655.

Gould WP and McGuire RG (2000). Hot water treatment and insecticidal coating for disinfecting limes of mealybugs (Hemiptera: Pseudococcidae). *J. Econ. Ento.* 93: 1017–1020.

Grafton-Cardwell EE, Millar JG, O'Connell NV, and Hanks LM (2000). Sex pheromone of yellow scale *Aonidiella citrina* (Homoptera: Diaspididae): evaluation as an IPM tactic. *J. Agric. Urban Ent.* 17: 75–88.

Gravena S, Coletti A, and Yamamoto PT (1992). Influence of green cover with *Ageratun conyzoides* and *Eupatorium pauciflorum* on predatory and phytophagous mites in citrus. *Proc. Int. Soc. Citriculture,* vol. 3, pp. 1259–1262.

Grout TG and Richards GI (1992). Integrated pest management of citrus thrips (Thysanoptera: Thripidae) in the Eastern Cape Province of South Africa. *Proc. Int. Soc. Citriculture,* vol. 3, pp. 920–923.

Grove T, Giliomee JH, and Pringle KL (2000). Efficiency of colored sticky traps for citrus thrips *Scirtothrips aurantii* Faure (Thysanoptera: Thripidae) in mango ecosystems of South Africa. *Fruits* 55: 253–258.

Gurung Anita, Sarkar TK, Mukhopadhyay S, and Ghosh MR (1993). Incidence of the vectors of tristeza virus in Darjeeling hills. *J. Ent. Res* 17 (2): 129–136.

Hallman GJ and Rene-Martinez L (2001). Ionizing irradiation quarantine treatment against Mexican fruit fly (Diptera: Tephritidae) in citrus fruits. *Post-harvest Biol.Tech.* 23: 71–77.

Hart WG (1978). Some biological control successes in the Southern United States. *Proc. Int. Soc. Citriculture,* vol. 2, pp. 154–156.

Inamullah Khan and Morse JG (1999). Field evaluation of *Chrysoperla* Spp. as predator of citrus thrips. *Sarhad J. Agric.* 15: 607–610.

Jeppson LR (1977). Binomics and control of mites attacking citrus. *Proc. Int. Soc. Citriculture,* vol. 2, pp. 445–451.

Jones VP and Morse JG (1984). A synthesis of temperature dependent developmental studies with the citrus red mite *Panonychus citri* (McGregor) (Aceri: Tetranychidae). *Flor. Entol.* 67: 213–221.

Katsoyannos BI, Papadopoulos NT, Heath RR, Hendrichs J, and Kouloussis NA (1999). Evaluation of synthetic food-based attractants for female Mediterranean fruit flies (Diptera: Tephritidae) in McPhail type traps. *J. Appl. Ent.* 123: 607–612.

Kim KyuChin, Choi DuckSoo, Kim KC and Choi DS (2000). Natural enemies of citrus red mite *Panonychus citri* McGregor, and seasonal occurrence of major predators on Yuzu tree (*Citrus junos*). *Korean J. Appl. Ent.* 39: 13–19.

Knapp JL and Browning HW (1989). Citrus black fly: Management in commercial orchards. *Citrus Industry* 70: 50–51.

Lai BangHai, Wu QiNeng, Lei BH, and Wu QN (2000). Trials on the effects of abamectin-petroleum in controlling citrus leaf miner. *Plant Protection* 26: 37–38.

Lama TK, Regmi C and Aubert B (1988). Distribution of citrus greening disease vector (*Diaphorina citri* Kuwayama) in Nepal, and to establishing biological control against it. *Proc. 10th Conf. IOCV* (eds. LW Timmer, SM Garnsey and LV Navarro). Univ. Calif. Riverside, CA, pp. 255–257.

Legaspi JC, French JV, Zuniga AG, and Legaspi BC Jr (2001). Population dynamics of the citrus leaf miner *Phyllocnistis citrella* (Lepidoptera: Phyllonictidae) and its natural enemies in Texas and Mexico. *Biol. Control* 21: 84–90.

Lei-BangHai, Wu-QiNeng, Lei BH and Wu QN (2000). Trials on the effects of abamectin-petroleum in controlling *Citrus* leafminer. *Plant Proct.* 26 (6): 37–38.

Lei-BangHai, Wu-QiNeng, Lei BH and Wu QN (2000). Trials on the effects of abamectin-petroleum in controlling *Citrus* leafminer. *Plant Proct.* 26 (6): 37–38.

Lei HuiDe, Lin BangMao, Ran Chun, Ling HongJun, Zhang QuanBing, Lei HD et al. (2000). Experiment of control of citrus red mite by 20% Fumanlang emulsion. *South China Fruits* 29: 21.

Li HongYun, Hu JunHua, Lei HuiDe, Ran Chun, Lin BangMao, Zhang QuanBing et al. (2001). Preliminary Report on the resistance of citrus red mite (*Panonynchus citri* McGr.) of citrus germplasm. *South China Fruits* 30: 18.

Lu ShengJin and Lu SJ (2000). Experiment of control of citrus rust mite by enhanced abamectin. *South China Fruits* 29: 25.

Luo ZhiYi, Zhou ChanMin, Luo JY, and Zhou CM (2001). Record of the citrus white fly in China. *South China Fruits* 30: 14–16.

Malleshaiah, Rajagopal BK and Gowda KNM (2000). Ecology of citrus mealy bug *Planococcus citri* (Risso.) (Hemiptera: Pseudococcidae). *Crop Research Hisar* 20: 130–133.

Mansanet V, Sanz JV, Izquierdo JI, and Puiggros-Bove JM (1999). Imidacloprid: a new strategy for controlling the citrus leaf miner (*Phyllocnistis citrella*) in Spain. Special issue: Imidacloprid. *Pflanzenschutz Nachrichten Bayer* 52: 360–373.

Mc Laren IW (1978). Biological control of citrus scale pest. *Proc. Int. Soc. Citriculture,* vol. 2, pp. 147–149.

Ming-Dau Huang, Siu-Wui Mai, Shu-xin Li and Jin Situ (1981). Biological control of citrus red mite *Panonychus citri* (McG.) in Guangdong Province. *Proc. Int. Soc. Citriculture,* vol. 2, pp. 643–645.

Miranda MA, Alonso R, and Alemany A (2001). Field evaluation of Med-fly (Diptera: Tephritidae) female attractants in a Mediterranean agro-system (Baiearic island, Spain). *J. Appl. Ent.* 125: 333–339.

Moreno DS and Kennett CE (1981). Monitoring insect pest populations by trapping in California citrus orchards. *Proc. Int. Soc. Citriculture,* vol. 2, pp. 684–687.

Morse JG (1986). Arthropod management in California citrus nurseries *In*: Int. Soc. Citrus Nurserymen 2nd World Congress (ed. DJ Gumpf), Riverside, California CA, pp. 193–197.

Morse JG and Brawner OL (1986). Toxicity of pesticides to *Scirtothrips citri* (Thysanoptera: Thripidae) and implications to resistance management. *J. Econ. Entomol.* 79: 565–570.

Nishida R, Shelly TE, Whitter TS and Kaneshiro KY (2000). Alpha-Copaene, a potential rendezvous cue for the Mediterranean fruit fly *Ceratitis capitata*? *J. Chem. Ecol.* 26: 87–100.

Papacek D and Smith D (1992). Integrated pest management of citrus in Queensland, Australia. *Proc. Int. Soc. Citriculture,* vol. 3, pp. 973–977.

Pehrson JE, Flaherty DL, O'Connell NV, and Phillips PA (eds), (1984). Integrated Management of Pests for Citrus. Univ. Calif. Div. Ag. Sci. Publ. 3303, 144 pp.

Qing Tang Yu (1990). On the parasitic complex of *Diaphorina citri* Kuwayama (Homoptera: Psyllidae) in Asian, Pacific and other Areas. *In*: Rehabilitation of Citrus Industry in the Asia Pacific Region (eds. B Aubert, S Tontyaporn, D Buangsuwon). UNDP-FAO Regional Project RAS/86/022: 240–245.

Raciti E, Barraco D and Conti F (2001). Biological control of citrus mealy bug. *Informatore Agrario* 57: 49–54.

Rezk HA (2001). The false spider mite, *Brevipalus obovatus* Donnadieu (Acari: Tnuipalpidae): host related biology, seasonal abundance and control. *Proc.10th Int. Cong. Acarology* (eds. RB Halliday, DE Walter, HC Proctor, RA Norton, and MJ Colloff), pp. 291–294.

Riehl LA (1981). Fundamental consideration and current development in production and use of petroleum oils. *Proc. Int. Soc. Citriculture,* vol. 2, pp. 602–607.

Samways MJ (1990). Biogeography and monitoring of outbreaks of the African citrus psyllid *Trioza erytreae* (Del Guercio). *In*: Rehabilitation of Citrus Industry in the Asia Pacific Region (eds. B. Aubert, S. Tontyapom, and D. Buangsuwon). UNDP-FAO Regional Project RAS/86/022: 188–197.

Samways MJ, Tate BA, and Mardoch E (1987). Population levels of adult citrus thrips (Thysanoptera: Thripidae) fecundity. *J. Econ. Entomol.* 84 (4): 1169–1174.

Spiccarelli R, Battaglia D, and Tranfaglia A (1992). Biological control of *Planococcus citri* (Risso) by *Leptomastix dactylopii* Howard in Citrus Grove in Metapontum area. *Proc. Int. Soc. Citriculture,* vol. 3, pp. 966–967.

Tagaki K (1981). Evaluation of the parasitoids of Latent Pests with a sticky suction trap. *Proc. Int. Soc. Citriculture,* vol. 2, pp. 627–629.

Tanigoshi LK (1981). Binomics and pest status of the citrus thrips *Scirtothrips citri* (Thysanoptera: Thripidae). *Proc. Int. Soc. Citriculture,* vol. 2, pp. 677–683.

Tanigoshi LK and Moreno DS (1981). Traps for monitoring populations of the citrus thrips *Scirtothrips citri* (Thysanoptera: Thripidae). *Can. Ent.* 113: 9–12.

Tanigoshi LK, Nishio-Wong JY, and Fargerlund JY (1984). *Euseius hibisci*: its control of citrus thrips in Southern California citrus orchards. *In*: Acarology VI, vol 2 (eds. DA Griffiths and CE Owan), Ellis Horwood Ltd., Chechester, UK.

Tanigoshi LK, Fargerlund J, Nishi-Wong JY, and Griffiths HJ (1985). Biological control of citrus thrips *Scirtothrips citri* (Thysanoptera: Thripidae) in Southern California citrus groves. *Environ. Entomol.* 14: 733–741.

Trevizoli D and Gravena S (1984). Management strategies for *Phyllocoptruta oleivora* (Ashm.) and their effects on other phytophagous species and natural enemies. *Proc. Int. Soc. Citriculture,* vol. 2, pp. 455–456.

Ujiye T (2000). Biology and the control of the citrus leaf-miner *Phyllocnistis citrella* Stainton (Lepidoptera: Phyllonyctidae) in Japan. *JARQ* 34: 167–173.

Wakgari WM and Giliomee JH (2000). Fecundity, fertility and phenology of white wax scale *Ceroplastes destructor* Newstead (Hemiptera: Coccidae) on *Citrus* and *Syzygium* in South Africa. *African Entomology* 8: 233–242.

Wakgari WM and Giliomee JH (2000). Fecundity, fertility and phenology of white wax scale, *Ceroplastes destructor* Newstead (Hemiptera: Coccidae) on citrus and *Syzygium* in South Africa. *African Entomology* 8 (2): 233–242.

Wang LianDe, You MinSheng and Wu Qing (1999). Damage of citrus leaf miner to citrus and its economic threshold. *Chinese J. Appl. Ecol.* 10: 457–460.

Wang Lian De, You Min Sheng and Wu Qing (1999). Damage of citrus leaf miner to citrus and its economic threshold. *Chinese J. Appl. Ecol.* 10: 457–460.

Watanabe MA, Tambasco FJ, Costa VA, Nardo EAB de, Facanali R and de Nardo EAB (2000). Population dynamics of some armored scales in citrus trees in different Sao Paulo State localities, Brazil. *Laranja* 21: 49–64.

Ying XueYang, Lin Ming, Xu WenZhen, Huang MingJie, Pan XingJie, Ying Xy et al. (2000). Effects of Dithane M 45 on the control of citrus rust mite and on fruit appearance and quality. *South China Fruits* 29: 14.

Yokomi R (1992). Potential for biological control of *Toxoptera citricidus* (Kirkaldy). *In*: Citrus Tristeza Virus and *Toxoptera citricidus* in Central America: Development of Management Strategies and Use of Biotechnology for Control (eds. Ramon Lastra, Richard Lee, Mario Roca-Peria, CL Nibblet, F Ochoa, SM Garnsey, and RK Yokomi) Maracabo, Venezuela, pp. 194–198.

Zhang QuanBing, Lei HuiDe, Lin BangMao, Ran Chun, Li Hongyun, Tian WenHua et al. (2001). Experimental control of citrus leaf miner. *South China Fruits* 30: 17.

Zhang XingLong and Zhang XL (2000). The extra effective insecticides for control of pomelo scale. *South China Fruits* 29: 23.

Appendixes

I

Some Commercial Cultivars of Mandarins, Tangors, Tangelo and Hybrids

1. Amanatsu, 2. Be-yue-ju, 3. Bendizao, 4. Beauty, 5. Biangan, 6. Bower, 7. Ciaculli, 8. Coorg, 9. Cravo, 10. Cuipigan, 11. Dahongpao, 12. Daoxiayeju, 13. Darjeeling, 14. Encore, 15. Fairchild, 16. Fortuna, 17. Fremont, 18. Fuju, 19. Goushigan, 20. Hai lai tangor, 21. Hassaku, 22. Hongpisuanju, 23. Imperator, 24. Iyokan, 25. Ju-long-ka, 26. Kara, 27. Khasi, 28. King, 29. Kinnow, 30. Kiyom tangor, 31. Ladu, 32. Langkat, 33. Manju, 34. Miyauchi, 35. Nagpur, 36. Nanfengmiju, 37. Natal, 38. Nianju, 39. Nova, 40. Onesco, 41. Pan American, 42. Parson's special, 43. Petyala, 44. Ponkan, 45. Queen, 46. Shantousuanju, 47. Shiju, 48. Som-pon, 49. Son-keo-wan, 50. Sunki, 51. Tankan, 52. Wilking, 53. Willow leaf, 54. Yuanjigjinagan, 55. Zanju

Clementine: 1. Ain Taoujat, 2. Azemmour, 3. Cadoux, 4. Cartenoire, 5. Commune SRA 63, 6. Commune SRA 93, 7. Corsica 2, Courtois, 8. Fine, G.P., 9. Janvier, 10. Larache, 11. Monreal, 12. Monreal GA 137, 13. Muska, 14. Nour, 15. Nules, 16. Nules SRA 334, 17. Oroval Marisol, 18. Oroval SRA 340, 19. Sidi Aissa, 20. 2000 SRA 272. 21. Tardive 2

Satsuma: 1. Aoe, 2. Aoshima, 3. Dobashebeni, 4. Guchi No. 4, 5. Hashimoto, 6. Hayashi, 7. Ichifumi, 8. Ikeda, 9. Imada, 10. Inaba, 11. Ishikama, 12. Ishitsuka, 13. Kamci, 14. Kidaguchi, 15. Kunou, 16. Matsuyama, 17. Mitake, 18. Miyamoto, 19. Miyagawa, 20. Morita, 21. Mukaiyama, 22. Nankan No. 4, 23. Nankan No. 20, 24. Okitsu, 25. Oomura, 26. Ooura, 27. Owari, 28. Saigon, 29. Seto, 30. Suruga, 31. Tachima, 32. Takabashi, 33. Tanimoto, 34. Tukumori, 35. Tomono, 36. Ueno, 37. Wakiyama, 38. Yamakawa No. 3, 39. Yonezawa

Tangor (mandarin sweet orange): Ellandale, Ortanique, Dweet, Murcott, Temple

Tangelo (mandarin grapefruit): Mapo, Minneola, Orlando, Sampson

II

Some Commercial Varieties of Sweet Oranges and Hybrids

1. Anliucheng, 2. Barao, 3. Belladona, 4. Bintangeheng, 5. Blood Red, 6. Calderon, 7. Calendra punchosa, 8. Chini, 9. Delta, 10. Diller, 11. Do Ceu, 12. Double fina, 13. Edangen No. 8, 14. Fohlha Murcha, 15. Gailiangcheng, 16. Hamlin, 17. Hiroshima, 18. HuGzhaucheng, 19. Jaffa, 20. Jingcheng, 21. Junar, 22. Koethen, 23. Kona, 24. Lima 25. Minera, 26. Luchen, 27. Madam(e)Vinous, 28. Malta, 29. Maltaise, 30. Mediterranean, 31. Midnight, 32. Moro, 33. Mosambi, 34. Moscato, Olikda, Ovale, 35. Natal, 36. Natsidaidai, 37. Navalina, 38. Navel (Bahia), 39. Navelate, 40. Newhall, 41. Page, 42. Parson Brown, 43. Pera, 44. Pineapple, 45. Recalate, 46. Robertson, 47. Royal temprana, 48. Ruby blood, 49. Saangre, 50. Salustiana, 51. Sanguine, 52. Sanguinello, 53. San Marino, 54. Santomera, 55. Satgudi, 56. Serra D'Agua, 57. Shirayanagi, 58. Suzuki, 59. Taoyecheng, 60. Tarocco, 61. Tarocco blood, 62. Tripoli, 63. Valencia (Cambell, Frost, Late, Rohde Red, Seedless, Temprana), 64. Verna, 65. Verna Peret, 66. Waialus, 67. Washington, 68. Washington sanguine, 69. Westin, 70. Xinhuicheng, 71. Xuegan

Citrange (*Citrus sinensis* *Poncirus trifoliata*)
1. Carrizo, 2. Mautauban, 3. Rusk, 4. Troyer

III

Some Commercial Varieties of Grapefruit (*C. paradisi* Macf.) and Hybrids

1. CRC 343. 2. Duncan, 3. Flame, 4. Foster Pink, 5. Ganganagar Red, 6. Marsh Seedless, 7. Pernambuco, 8. Redblush, 9. Rio Red, 10. Ruby, 11. Sharanpur Special, 12. Shamber, 13. Star Ruby

Citrumelo (*C. paradisi P. trifoliata)*

IV

Some Commercial Varieties of Lemon (*C. limon* (L.) Burm. f.)

1. Femminello Siracusano, 2. Eureka Allen, 3. Eureka Cascade, 4. Eureka Cook, 5. Eureka Corona, 6. Eureka Frost, 7. Hill, 8. Interdonato, 9. Kara, 10. Lisbon, 11. Libone Calier, 12. Lisbon Frost, 13. Lisbon Prior, 14. Molla Mehmet, 15. Monachello, 16. Royal, 17. Santa Teresa

V

Some Commercial Varieties of Limes and Sweet Limes

Citrus aurantifolia (Christm.) Swing
1. Goal Nemu, 2. Kagzi, 3. Kagzikalan, 4. Mexican, 5. Sans Epines

Citrus latifolia Tan.
1. Bearss, 2. Persian, 3. Tahiti

Citrus limonia Osb.
Rangpur lime

Sweet Lime (*C. limettoides* Tan.): Palestine

VI

Some Commercial Varieties of Rough Lemon (*C. jambhiri* Lush)

1. Estes, 2. Florida, 3. Honglimeng, 4. Jatti Khatti, 5. Limeng, 6. Seti Jamir, 7. Tuningmeng

VII

Some Commercial Varieties of Sour Orange (*C. aurantium* L.) and Hybrid

1. Bangan, 2. Bergamota, 3. Bittersweet, 4. Daidai, 5. Goutoucheng, 6. Smooth Seville, 7. Xiaohongcheng, 8. Xingshanchuancheng, 9. Zuluan

VIII

Some Commercial Varieties of Trifoliate Orange (*Poncirus trifoliata* (L.) Raf.

1. Large flower, 2. Large leaf, 3. Rubidoux, 4. Small leaf. 5. Small flower

IX

Composition of Requied Ingredients of "Soilmix"

Peat sand: 1/2 peat moss/substitute: 1/2
Superphosphate: 1.7 kg/m^3 Dolomite 2.25 kg/m^3; Calcium carbonate: 1.0 kg/m^3

Micronutrients
$CuSO_4$, $5H_2O$: 8 kg/m^3; $ZnSO_4$ (36% Zn): 34 g/m^3; $MnSO_4$ (28% Mn): 37 g/m^3; $FeSO_4$, $7H_2O$: 0.25 g/m^3; H_3BO_4: 0.75 g/m^3

X

Adaptability of Traditional and Conventional Rootstocks to Various Soil Conditions (soil structure, drought, waterlogging, calcium, chloride, and boron)

Soil/water constraints	Well adapted Easy to grow	Moderately adapted Average growth	Nonadaptable Difficult to grow
Sandy soil	*Citrus macrophylla* *Citrus volkameriana* Rangpur lime	Common sour orange Gou Tou sour orange Citrumelo 4475 Citrumelo sacation Cleopetra mandarin Fuzhu mandarin	Citranges Troyer, Carrizo Trifoliate orange Trifoliate orange- Flying dragon
Loamy/clay	Gou Tou sour orange Carrizo citrange Trifoliate orange- Flying dragon Fuzhu mandarin	*Citrus macrophylla* Common sour orange Troyer citrange Cleopetra mandarin Orlando tangelo Rangpur lime	Citrumelo 4475 Citrumelo sacation *Citrus volkameriana*
Tolerance to drought	*Citrus macrophylla* *Citrus volkameriana* Rangpur lime	Common sour orange Gou Tou sour orange Carrizo citrange Troyer citrange Citrumelo 4475 Cleopetra mandarin Orlando tangelo Fuzhu mandarin	Trifoliate orange Trifoliate orange- Flying dragon
Calcareous soil	*Citrus macrophylla* Cleopetra madarin	Common sour orange Gou Tou sour orange	Carrizo citrange Troyer citrange

Calcareous soil	*Citrus volkameriana* Trifoliate orange-Flying dragon Orlando tangelo Rangpur lime		Citrumelo 4475 Citrumelo sacation Trifoliate orange
Acidic soil	Carrizo citrange Troyer citrange Citrumelo 4475 Citrumelo sacation Trifoliate orange Trifoliate orange-Flying dragon Fuzhu mandarin	Common sour orange Gou Tou sour orange *Citrus volkameriana* Cleopetra mandarin Rangpur lime	*Citrus macrophylla*
Chlorides	*Citrus macrophylla* *Saverinia buxiflora* Rangpur lime	Common sour orange Gou Tou sour orange Citrumelo 4475 Citrumelo sacation *Citrus volkameriana* Cleopetra mandarin Orlando tangelo Fuzhu mandarin	Carrizo citrange Troyer citrange Trifoliate orange Trifoliate orange-Flying dragon
Boron	*Citrus macrophylla* *Saverinia buxiflora*	Common sour orange Gou Tou sour orange Carrizo citrange Troyer citrange Citrumelo 4475 Cleopetra mandarin Orlando tangelo Fuzhu mandarin Rangpur lime	Trifoliate orange Trifoliate orange-Flying dragon

Source: Aubert and Vullin, 1998

XI

Tolerance of Some Important Rootstocks to Certain Climatic Conditions and Diseases

Rootstock	1	2	3	4	5	6	7	8	9	10	11	12
Rough lemon	GT	GT	PT	S	PT	T	T	T	S	S	HT	H
Volkamer lemon	GT	GT	PT	S	PT	T	T	T	S	S	HT	H
Rangpur lime	?	GT	PT	S	PT	T	S	S	S	S	HT	H
Palestine Sweet orange	GT	GT	PT	S	PT	T	S	S	S	S	I	H
Sour orange	I	I	GT	T	GT	S	T	T	S	S	HT	I
Cleopetra Mandarin	PT	I	GT	T	GT	T	T	T	S	S	I	H
Sweet orange	PT	PT	I	S	GT	T	T	T	S	S	L	I
Trifoliate orange	I	PT	GT	R	PT	R	S	T	S	R	P	L
Carrizo citrange	P	G	I	T	PT	T	S	T	T	T	L	H
Troyer citrange	G	I	G	R	?	T	T	T	S	R	?	I
Swingle citrumelo	?	G	I	T	?	T	S	T	S	T	L	H
Grapefruit	?	PT	I	S	?	T	T	T	S	S	L	H

1—Flood tolerance, 2—Drought tolerance, 3—Frost tolerance, 4—Phytophthora tolerance, 5—Blight tolerance, 6—Tristeza tolerance, 7—Exocortis tolerance, 8—Xyloporosis tolerance, 9—Burrowing nematode tolerance, 10—Citrus nematode tolerance, 11—Calcium tolerance, 12—Tree vigor

1. GT—Good tolerance, 2. H—High, 3. HT—High tolerance, 4. I—Intermediate, 5. L—Low, 6. PT—Poor tolerance, 7. R—Resistant, 7. S—Susceptible, 8. T—Tolerant, 9. ?—Inadequate information/Rating unknown

Source: Castle et al., 1987

XII

Use of Different Rootstock Scionic Combinations in Various Countries

Country	Scionic varieties	Rootstock used
Brazil		
	Pera sweet orange (a local orange Latee variety), Valencia, Bahia (Navel), Hamlin, Ponkan, Cravo, Murcott, Willow leaf tangerines, Mandarin, Tahiti lemon, Mexican lime	Rangpur lime, sweet orange, Trifoliate orange and its hybrids, Cleopetra mandarin
Argentina		
	Valencia late, Calderon, Navels, Hamlin, Common oranges (local variety), Genova and Ureka lemons, Willow leaf Tangerines, Satsuma, Campeona, Beauty, Malvasis, Ellendale Mandarins, Marsh seedless, Ruby and Duncan grapefruits	Trifoliate orange, Rangpur lime, Common sweet orange Florida Rough lemon, Troyer citrange, Cleopetra mandarin
Peru		
	Washington Navel, Valencia, Hamlin, Criolla orange (local) Mexican lime, Dancy, Satsuma, Ponkan mandarin, Marsh grapefruit, etc.	Categorization not available
Venezuela		
	Valencia, Washington Navel, Pineapple oranges, Dancy mandarin, Marsh seedless, Thompson, Foster, Duncan grapefruit, Mexican lime	Sour orange
Uruguay		
	Valencia, Navels, Hamlin, Ellendale, Willow leaf, Satsuma, Malaquina, Malvasis mandarins,	Trifoliate orange

	Ruby Red, Star Ruby grapefruit, True lemon (Genoa type), etc.	
Australia	Washington Navel, Leng Navel, Valencia late, Joppa (mid-season orange), Hamlin, Pine apple, Eureka selections, Lisbon selections, Marsh, Wheeny, Thompson, Ruby Red, Foster, Imperial, Satsuma (silver hill), Unshiu, Emperor, Ellendale, Kara, Seminole tangelo, Minneola tangelo	Trifoliate orange for most scions of oranges and lemons except Eureka lemon. Troyer and Carrizo citranges for oranges and grapefruits

XIII

Citrus Insect Pests of Various Countries and Their Preferred Citrus Hosts

Pest hosts		Parts infested	Common citrus
I.	Thysanoptera Thripidae		
1.	*Heliothrips haemorrhoidalis* (Black tea thrips)	Leaves, fruits	orange
2.	*Scirtothrips citri* (Moulton) (Citrus thrips)	Leaves, flowers, fruits	some varieties
3.	*Scirtothrips aurantii* Faure (South African citrus thrips)	Twigs, fruits	some varieties
4.	*Frankliniella occidentalis* (Pergande) (Western flower thrips)	Flowers	some varieties
5.	*Thrips tabaci* Lind (Onion thrips)	Leaves, flowers, fruits	some varieties
II.	Heteroptera		
6.	*Calocoris trivialis* (Costa).	Flowers	orange, lemon
7.	*Rhynchocoris humeralis* (Thumb.) (Green stink bug)	Leaves, twigs, branches	some varieties
8.	*Leptoglossus phyllopus* (L). (Leaf-footed bug)	Leaves	some varieties
III.	Homoptera		
9.	*Toxoptera aurantii* (B. de F.) (Black citrus aphid)	Leaves	all common citrus varieties
10.	*Toxoptera citricida* (Kirkldy). (Brown citrus aphid)	Leaves	all common citrus varieties
11.	*Myzus persicae* (Sulzer.) (Green peach aphid)	Leaves	all common citrus varieties
12.	*Aulacorthium solani* Kalt.	Leaves	all common citrus varieties
13.	*Aphis gossypii* Glover (Cotton aphid)	Leaves	occasionally on some varieties
14.	*Aphis craccivora* Koch	Leaves	occasionally on some varieties
15.	*Aphis solani* (Kattarbach).	Leaves	occasionally on some varieties
16.	*Aphis citricola* van der Goot (Green citrus aphid)	Leaves	occasionally on some varieties

17.	*Aphis spiraecola* Patch (Spirea aphid)	Leaves	occasionally on some varieties
18.	*Sinomegaura citricola* van der Goot	Leaves	occasionally on some varieties
19.	*Empoasca smithi* (Australian leafhopper)	Leaves	occasionally on some species
20.	*Neoaleturus hematoceps* (Leafhopper)	Leaves	occasionally on some species
21.	*Neoalturus tenellus*	Leaves	occasionally on some species
22.	*Diaphorina citri* Kuw.	Leaves, twigs	on most species
23.	*Trioza erytreae* Del Guercio	Leaves, twigs	on most species
24.	*Dialeurodes citri* (Ashm.) (Citrus whitefly)	Leaves	orange
25.	*Dialeurodes citrifolii* (Morg.) (Cloudy winged whitefly)	Leaves	occasionally on some varieties
26.	*Parabemisia myricae* (Kuwana) (Bayberry whitefly)	Leaves	occasionally on some varieties
27.	*Aleurothrixus floccosus* (Muskell) (Flocculent (wooly) whitefly)	Leaves	occasionally on some varieties
28.	*Aleurocanthus woglumi* Ashby (Citrus black fly)	Leaves	orange, lemon, grapefruit, lime
29.	*Aleurocanthus citriperdus* (Quaintence et Baker) (Citrus black fly)	Leaves	orange, lemon, grapefruit, lime
30.	*Aleurocanthus spiniferus* Quaint. (Orange spiny whitefly)	Leaves	on some varieties
31.	*Orchamoplatus citri* (Takahashi). (Japan citrus whitefly)	Leaves	occasionally on some varieties
32.	*Aonidiella aurantii* (Maskell) (California red scale) fruits	Leaves, twigs, branches	all common varieties
33.	*Aonidiella citrina* (Coquillat) (Yellow scale)	Leaves, twigs, branches, fruits	all common varieties
34.	*Lepidosaphes beckii* (Newman) (Purple scale)	Leaves, twigs, branches, fruits	all common varieties
35.	*Lepidosaphes* (= *Insulaspis*) *Gloverii* (Packard) (Glover scale)	Leaves, twigs, branches, fruits	all common varieties
36.	*Aspidiotus nerii* Bouche (White scale)	Leaves, twigs, branches, fruits	orange, lemon
37.	*Parlatoria zizyphi* (Lucas) (Black scale)	Leaves, twigs, branches, fruits	orange, lemon, tangerine citron
38.	*Parlatoria pergandii* (Chaff scale)	Comstock Leaves, twigs, branches fruits	occasionally on some varieties
39.	*Chrysomphalus dictyospermi* (Morg.) (Dictyospermi (fern) scale, (Spanish Red scale)	Leaves, twigs, branches	all common varieties
40.	*Chrysomphalus ficus* Ashmead (Purple scale)	Leaves, twigs, branches	all common varieties

41.	*Pinnaspis aspidistrae* (Sign.) (Aspidistra scale)	Leave, twigs, branches	on some varieties
42.	*Mytilococcus beckii* (Newman) (Mussel scale)	Leaves, twigs, branches	all common varieties
43.	*Mytilococcus gloverii* (Pachard) (Citrus long scale)	Leaves, twigs, branches	all common varieties
44.	*Unaspis citri* (Comstock) (Citrus snow scale)	Leaves, twigs, branches	some varieties
45.	*Unaspis yanonensis* (Kuwana) (Arrowhead scale)	Leaves, twigs, branches	some varieties
46.	*Selenaspis articulatus* Morgan (Brown scale)	Leaves, twigs, branches	some varieties
47.	*Icerya purchasi* Maskell (Cottony cushion scale)	Leaves, twigs, branches	all common varieties
48.	*Labyoproctus polei* (Green) (Mealy bugs)	Leaves, twigs, branches	some varieties
49.	*Saissetia oleae* (Olivier) (Mediterranean Black scale)	Leaves, twigs, branches	orange, lemon, sour orange, tangerine
50.	*Saissetia hemisphaerica* Targ. (Hard brown scale, coffee helmet scale)	Leaves, twigs, branches	some varieties
51.	*Coccus hesperidum* L. (Brown soft scale)	Leaves, twigs, branches	orange, lemon, sour orange, tangerine
52.	*Coccus pseudomagnoliarum* (Kuwana) (Citricola scale)	Leaves, twigs, branches	orange, lemon
53.	*Coccus viridis* Green (Coffee green scale)	Leaves, twigs, branches	some varieties
54.	*Ceroplastes rusci* L.	Leaves, twigs, branches	orange, tangerine
55.	*Ceroplastes floridensis* Comstock (Florida white wax scale)	Leaves, twigs, branches	orange, lemon, sour orange, tangerine
56.	*Ceroplastes destructor* Newstead (White wax scale)	Leaves, twigs, branches	some varieties
57.	*Ceroplastes rubens* Maskell (Pink wax scale)	Leaves, twigs, branches	some varieties
58.	*Ceroplastes sinensis* Del Guer. (Chinese wax scale)	Leaves, twigs, branches	some varieties
59.	*Pulvinaria floccifera* Westwood (White wax scale)	Leaves, twigs	orange, lemon
60.	*Planococcus citri* (Risso) (Citrus mealybug)	Leaves, twigs, branches	all common varieties
61.	*Pseudococcus adonidum* L. (Long tailed mealy bug)	Leaves, twigs, branches, fruits	orange, lemon
62.	*Pseudococcus longispinus* (Targioni-Tozzetti). (Long-tailed mealy bug)	Leaves, twigs, branches	some varieties
63.	*Pseudococcus citriculus* Green	Leaves, twigs, branches	some varieties

	(Citriculus mealy bug)		
64.	*Pseudococcus comstocki* Kuw. (Comstock mealybug)	Leaves, twigs, branches	some varieties
IV.	Coleoptera		
65.	*Carpophilus hemipterus* L.	Fruits	orange, sour orange
66.	*Epicometis hirtella* L.	Flowers	orange, lemon
67.	*Tropinita squalida* (Scop.)	Flowers	orange, lemon
68.	*Oxythyrea funesta* Poda	Flowers	orange, lemon
69.	*Anoplophora versteegi* (Rits.) (Trunk/shoot borer)	Trunk, shoots	some varieties
70.	*Chelidonium gibbicolle* White (Roundheaded borer)	Trunk, shoots	some varieties
71.	*Melanauster chinensis* Forster (Black and white citrus borer)	Trunk, shoots	some varieties
72.	*Otiorrhynchus cribricollis* Gyll.	Leaves	certain varieties
73.	*Pantomorus cervinus* Boh.	Leaves	certain varieties
V.	Lepidoptera		
74.	*Prays citri* Mill (Citrus flower moth)	Flowers, fruits	lemon, orange
75.	*Spodoptera littoralis* (Boisd.)	Leaves, twigs	lemon, orange
76.	*Apiphas postvittana* (Walker). (Light brown apple moth)	Leaves	orange
77.	*Archips rosanus* (L.). (Fruit tree leaf roller)	Leave	orange
78.	*Platynota stultana* Wasinghom (Omnivorus leaf roller)	Leaves	orange
79.	*Argyrotaenia citrana* (Ferneld). (Orange tortrix)	Leaves	orange
80.	*Ectomyelois ceratoniae* (Zeller). (Symbiosis with *Planococcus citri* Russo)	Fruits	orange
81.	*Cryptoblabes gnidiella* Mill. (Symbiosis with *Planococcus citri* Russo)	Fruits	orange
82.	*Phyllocnistis citrella* Stainton (Citrus leafminer)	Leaves, twigs	some varieties
83.	*Papilio crespontes* Lucas. (Citrus dog/ lemon caterpillar/black anise swallowtail)	Leaves	some varieties
84.	*Papilio anaetus* MaCleay. (Small citrus butterfly)	Flowers, leaves	some varieties
85.	*Papilio demoleus demoleus* L. (Citrus/lemon butterfly)	Leaves	some varieties
86.	*Papilio memom* L. (Pastor shallow tail)	Leaves	some varieties
87.	*Papilio polytes polytes* L	Leaves, flowers	some varieties

VI. Diptera		
88. *Ceratitis capitata* (Wied.) (Mediterranean fruit fly)	Fruits	common varieties
89. *Dacus caudetus* (Fabr.) (Oriental fruit fly)	Fruits	common varieties
90. *Dacus dorsalis* Herd. (Oriental fruit fly)	Fruits	common varieties
91. *Dacus tsuneonis* Hendel (Chinese citrus fruit fly)	Fruits	common varieties
92. *Anastrepha fraterculus* Wied. (South American fruit fly)	Fruits	common varieties
VII. Hymenoptera		
93. *Acromyrmex octospinosus* Reich (Leafcutting ant)	Leaves	some varieties
94. *Atta cephalotes* L. (Leafcutting ant)	Leaves	some varieties
95. *Atta sexdens* L. (Leafcutting ant)	Leaves	some varieties
96. *Solenopsis invicta* Buren (Fire ant, Citrus thrips)	Leaves	some varieties

XIV

Natural Enemies of *Toxoptera citricidus* (Yokomi, 1992)

PREDATORS

Coccinellidae

Chiomenes sexmacculata, Coccinella octopunctata, Coccinella repanda, Coccinella semptempunctata bruckii, Coccinella transversalis, Coelophora chinensis, Ceophora saucia, Leis dimidistus, Lemnia biplagiata, Lemnia swinhoei, Menochilus sexmaculatus, Platynaspadis maculosus, Pseudaspidimerusja ponensis, Scymnus frontalis quadripustulatus, Schymnus hilaris, Schymnus hoffmanni, Schymnus sp., *Synonucha grandis, Adalia bipunctata, Telsimia negra, Chillocorus kuwanae, Stethorus japonica, Hyperapsis japonica, Propylea japonica, Lemnia saucia, Hermonia octomaculata, Pentilea* sp., *Coccidophilus* sp., *Azya* sp., *Olla* sp., *Chilocorus* sp.

Chamaemyiidae

Leucopis sp., *Leucopis punticornis*

Syrphidae

Asarcina aegrota, Epistrophe balteata, Erystrophe horishana, Ishiodon scutellaris, Metasyrphus corollae, Paragus serratus, Paragus tibialis, Parxanthogramma nakamurae, Sphaerosphoria cylindira, Sphaerosphoria formosana, Sphaerosphoria javana, Syrphus aegrotus, Syrphus confrater, Syrphus horishanus, Syrphus serrarius, Syrphus vitripennis, Xanthogramma citrifasciatum

Chrysopidae

Ankylopteryx octopunctata, Mallada boninensis, Chrysopa formosana, Chrysopa sptempuntata

Hemerobiidae

Micromus timidus

PARASITES

Braconidae

Binidoxys scuterallis, Lysiphlebus sp., *Trioxys communis, Lysiphlebus japonica, Lipolexix gracilis, Aphidius colemani, Binodoxyx indicus, Lyshiphlebus delhiensis, Aphidius smithi*

Aphelindae

Aphelinus mali, Aphelinus sp.

Entomopathogens

Dueteromycetes
Paecilomyces fumosoroseus, Paecilomyces sp., *Beauveria basiana, Metarhizium anisopliae, Verticillium lecanii, Aschersonia* sp., *Cephalosporium* sp.

Entomophthorales

Empusa fresenii, Pandora neoapidis, Pandora nouryi, Pandora delphacis

XV

Parasites of *Diaphorina citri* Recorded in Asia-Pacific and Other Areas

Name of the parasite	Parasitic type	Region
Eulophidae		
1. *Tamarixia radiata* = *Tetrastichus radiata*	Primary	India, Reunion*, Saudi Arabia, Mauritius*, Nepal, Taiwan* Mainland China, Indonesia, Thailand
2. *Tetrastichus* sp.	Hyper—on TR & DA	Mainland China, Taiwan, Philippines
Encyrtidae		
3. *Diaphorencyrtus aligarhensis*	Primary	India, Vietnam
= *Aphydencirtus diaphorinae*		Taiwan, Comores island
= *Psyllaephagus diaphorinae*		Reunion, Philippines
= *Aphidencyrtus aligarhensis*		Main-land China, Indonesia
4. *Syrphophagus taiwanus*	Hyper—on TR & DA	Taiwan, Mainland China
5. *Ageniaspis* sp.	Hyper—on DA	Taiwan
6. *Cheiloneurus* sp.	Hyper—on DA	Taiwan
Signiphoridae		
7. *Chartocerus walkeri*	Hyper—on TR & DA	Taiwan, Mainland China
Pteromalidae		
8. *Pachyneuron concolor*	Hyper—on TA & DA	Taiwan
Aphelinidae		
9. *Coccophagus ceroplastae*	Hyper—on DA	Taiwan
10. *Coccophagus* sp.	Hyper—on DA	Taiwan
11. *Marietta leopardina* = *Marietta javensis*	Hyper—TA & DA DA Philippines	Taiwan
12. *Encarsia* sp.	Hyper—on TR & DA	Taiwan, Mainland China

TR = *T. radiata*; DA = *D. aligarhensis*
*Imported
Source: Qing, 1990

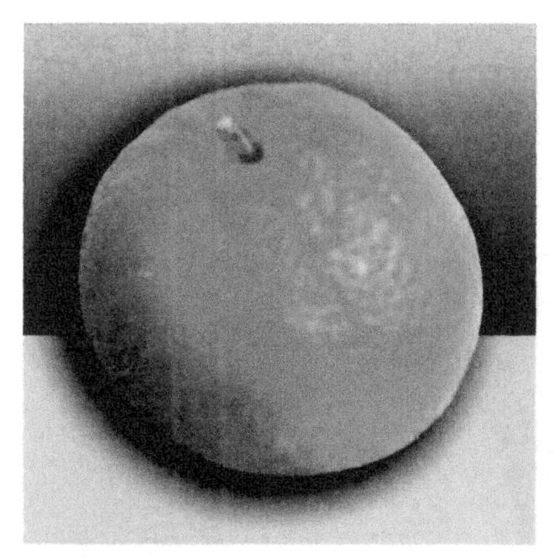

II.1 Bright orange colored seedless "Clementine" fruit (Israel)

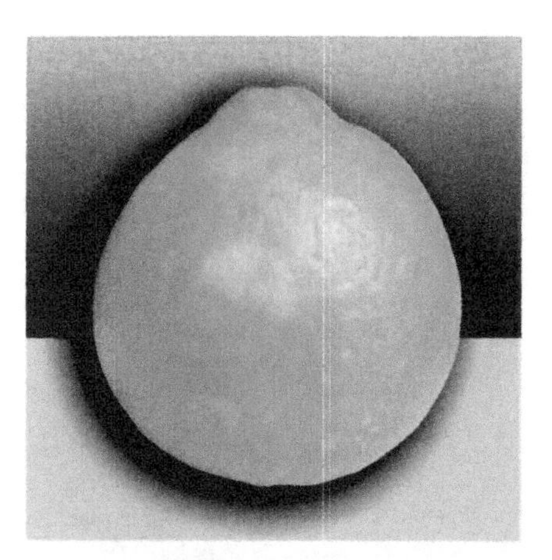

II.2 Fruit of a "Mnneola" mandarin orange (Israel), man-made hybrid between Dancy Tangerine and Duncan Grapefruit

II.3 "Shamouti" orange fruit (Israel), a bud mutation of Baladi orange

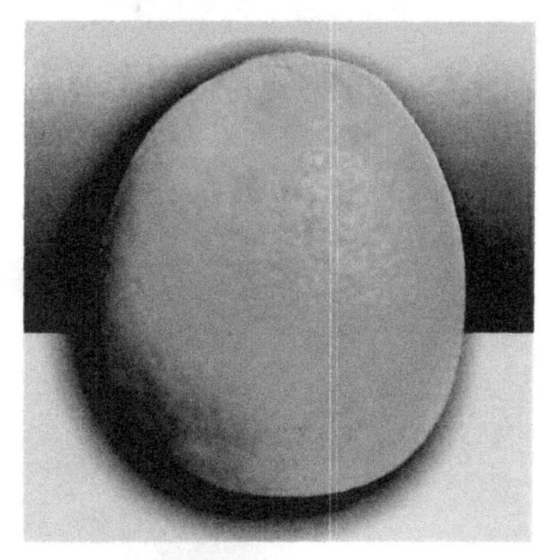

II.4 Fruit of "Navalena" (Spain), an artificially bred Navel orange

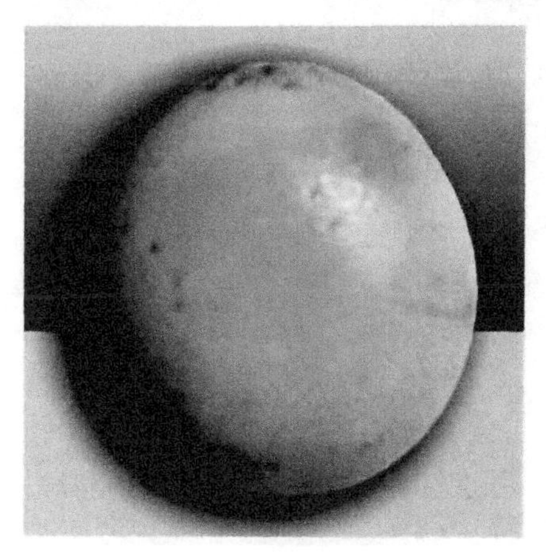

II.5 Fruit of "Flame" (USA), a Florida Red pigmented grapefruit

Plate I. Different types of *Citrus* fruits.

III.1 A type fruit of "Rough lemon" rootstock showing its rough and irregular surfce

VII.1 Typical leaf symptom of "Greening" in sweet orange indicator plant

Plate II.

VII.2 Typical vein thickening and stem pitting of "Tristeza virus (sever stain)" on Mexican lime indicator plant

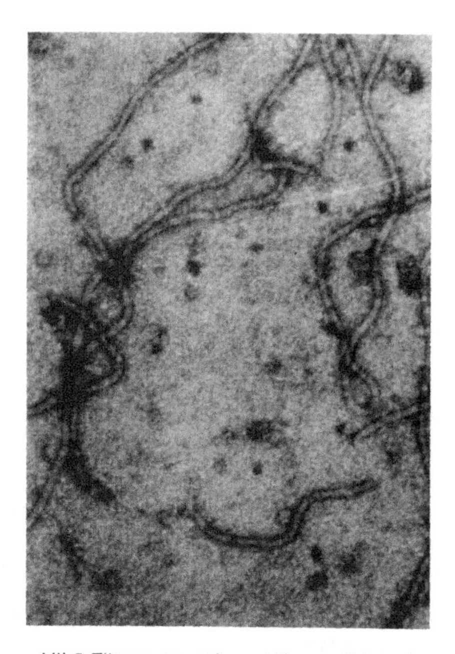

VII.3 Filamentous thread-like particles of "*Citrus* Tristeza Virus" as observed under an electron microscope (x1,60,00)
Courtesy: Professor YS Ahlawat, Advanced Center for Plant Virology, IARI, New Delhi

VII.4 Symptom of "Ring spot virus" infection on Darjeeling mandarin

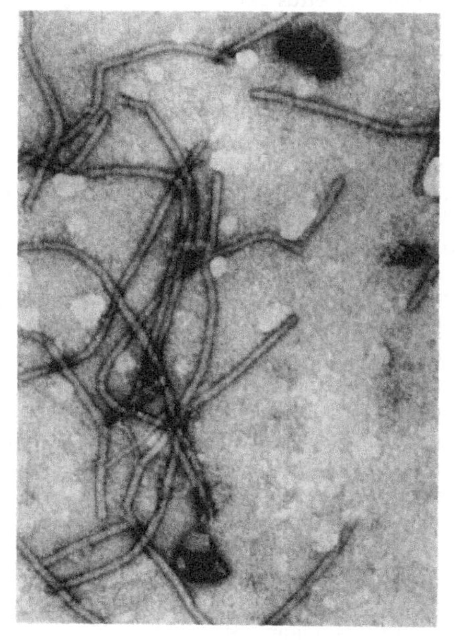

VII.5 Filamentous particles of "Citrus Ringspot Virus" as observed under an electronmicroscope.
Courtesy: Professor YS Ahlawat, Advanced Center for Plant Virology, IARI, New Delhi

Plate III.

Subject Index

Plant Species/Common Names Index

Author Index

For Product Safety Concerns and Information please contact our EU
representative GPSR@taylorandfrancis.com Taylor & Francis Verlag GmbH,
Kaufingerstraße 24, 80331 München, Germany

Printed and bound by CPI Group (UK) Ltd, Croydon, CR0 4YY
01/05/2025
01858546-0005